IET ENERGY ENGINEERING SERIES 161

Artificial Intelligence for Smarter Power Systems

Other volumes in this series:

Volume 1 **Power Circuit Breaker Theory and Design** C.H. Flurscheim (Editor)
Volume 4 **Industrial Microwave Heating** A.C. Metaxas and R.J. Meredith
Volume 7 **Insulators for High Voltages** J.S.T. Looms
Volume 8 **Variable Frequency AC Motor Drive Systems** D. Finney
Volume 10 **SF_6 Switchgear** H.M. Ryan and G.R. Jones
Volume 11 **Conduction and Induction Heating** E.J. Davies
Volume 13 **Statistical Techniques for High Voltage Engineering** W. Hauschild and W. Mosch
Volume 14 **Uninterruptible Power Supplies** J. Platts and J.D. St Aubyn (Editors)
Volume 15 **Digital Protection for Power Systems** A.T. Johns and S.K. Salman
Volume 16 **Electricity Economics and Planning** T.W. Berrie
Volume 18 **Vacuum Switchgear** A. Greenwood
Volume 19 **Electrical Safety: A Guide to Causes and Prevention of Hazards** J. Maxwell Adams
Volume 21 **Electricity Distribution Network Design, 2nd Edition** E. Lakervi and E.J. Holmes
Volume 22 **Artificial Intelligence Techniques in Power Systems** K. Warwick, A.O. Ekwue and R. Aggarwal (Editors)
Volume 24 **Power System Commissioning and Maintenance Practice** K. Harker
Volume 25 **Engineers' Handbook of Industrial Microwave Heating** R.J. Meredith
Volume 26 **Small Electric Motors** H. Moczala *et al.*
Volume 27 **AC–DC Power System Analysis** J. Arrillaga and B.C. Smith
Volume 29 **High Voltage Direct Current Transmission, 2nd Edition** J. Arrillaga
Volume 30 **Flexible AC Transmission Systems (FACTS)** Y.-H. Song (Editor)
Volume 31 **Embedded Generation** N. Jenkins *et al.*
Volume 32 **High Voltage Engineering and Testing, 2nd Edition** H.M. Ryan (Editor)
Volume 33 **Overvoltage Protection of Low-Voltage Systems, Revised Edition** P. Hasse
Volume 36 **Voltage Quality in Electrical Power Systems** J. Schlabbach *et al.*
Volume 37 **Electrical Steels for Rotating Machines** P. Beckley
Volume 38 **The Electric Car: Development and Future of Battery, Hybrid and Fuel-Cell Cars** M. Westbrook
Volume 39 **Power Systems Electromagnetic Transients Simulation** J. Arrillaga and N. Watson
Volume 40 **Advances in High Voltage Engineering** M. Haddad and D. Warne
Volume 41 **Electrical Operation of Electrostatic Precipitators** K. Parker
Volume 43 **Thermal Power Plant Simulation and Control** D. Flynn
Volume 44 **Economic Evaluation of Projects in the Electricity Supply Industry** H. Khatib
Volume 45 **Propulsion Systems for Hybrid Vehicles** J. Miller
Volume 46 **Distribution Switchgear** S. Stewart
Volume 47 **Protection of Electricity Distribution Networks, 2nd Edition** J. Gers and E. Holmes
Volume 48 **Wood Pole Overhead Lines** B. Wareing
Volume 49 **Electric Fuses, 3rd Edition** A. Wright and G. Newbery
Volume 50 **Wind Power Integration: Connection and System Operational Aspects** B. Fox *et al.*
Volume 51 **Short Circuit Currents** J. Schlabbach
Volume 52 **Nuclear Power** J. Wood
Volume 53 **Condition Assessment of High Voltage Insulation in Power System Equipment** R.E. James and Q. Su
Volume 55 **Local Energy: Distributed Generation of Heat and Power** J. Wood
Volume 56 **Condition Monitoring of Rotating Electrical Machines** P. Tavner, L. Ran, J. Penman and H. Sedding
Volume 57 **The Control Techniques Drives and Controls Handbook, 2nd Edition** B. Drury
Volume 58 **Lightning Protection** V. Cooray (Editor)
Volume 59 **Ultracapacitor Applications** J.M. Miller
Volume 62 **Lightning Electromagnetics** V. Cooray
Volume 63 **Energy Storage for Power Systems, 2nd Edition** A. Ter-Gazarian

Volume 65 **Protection of Electricity Distribution Networks, 3rd Edition** J. Gers
Volume 66 **High Voltage Engineering Testing, 3rd Edition** H. Ryan (Editor)
Volume 67 **Multicore Simulation of Power System Transients** F.M. Uriate
Volume 68 **Distribution System Analysis and Automation** J. Gers
Volume 69 **The Lightening Flash, 2nd Edition** V. Cooray (Editor)
Volume 70 **Economic Evaluation of Projects in the Electricity Supply Industry, 3rd Edition** H. Khatib
Volume 72 **Control Circuits in Power Electronics: Practical Issues in Design and Implementation** M. Castilla (Editor)
Volume 73 **Wide Area Monitoring, Protection and Control Systems: The Enabler for Smarter Grids** A. Vaccaro and A. Zobaa (Editors)
Volume 74 **Power Electronic Converters and Systems: Frontiers and Applications** A.M. Trzynadlowski (Editor)
Volume 75 **Power Distribution Automation** B. Das (Editor)
Volume 76 **Power System Stability: Modelling, Analysis and Control** A.A. Sallam and Om P. Malik
Volume 78 **Numerical Analysis of Power System Transients and Dynamics** A. Ametani (Editor)
Volume 79 **Vehicle-to-Grid: Linking Electric Vehicles to the Smart Grid** J. Lu and J. Hossain (Editors)
Volume 81 **Cyber-Physical-Social Systems and Constructs in Electric Power Engineering** S. Suryanarayanan, R. Roche and T.M. Hansen (Editors)
Volume 82 **Periodic Control of Power Electronic Converters** F. Blaabjerg, K. Zhou, D. Wang and Y. Yang
Volume 86 **Advances in Power System Modelling, Control and Stability Analysis** F. Milano (Editor)
Volume 87 **Cogeneration: Technologies, Optimisation and Implementation** C.A. Frangopoulos (Editor)
Volume 88 **Smarter Energy: From Smart Metering to the Smart Grid** H. Sun, N. Hatziargyriou, H.V. Poor, L. Carpanini and M.A. Sánchez Fornié (Editors)
Volume 89 **Hydrogen Production, Separation and Purification for Energy** A. Basile, F. Dalena, J. Tong and T.N. Veziroğlu (Editors)
Volume 90 **Clean Energy Microgrids** S. Obara and J. Morel (Editors)
Volume 91 **Fuzzy Logic Control in Energy Systems with Design Applications in MATLAB®/Simulink®** İ.H. Altaş
Volume 92 **Power Quality in Future Electrical Power Systems** A.F. Zobaa and S.H.E.A. Aleem (Editors)
Volume 93 **Cogeneration and District Energy Systems: Modelling, Analysis and Optimization** M.A. Rosen and S. Koohi-Fayegh
Volume 94 **Introduction to the Smart Grid: Concepts, Technologies and Evolution** S.K. Salman
Volume 95 **Communication, Control and Security Challenges for the Smart Grid** S.M. Muyeen and S. Rahman (Editors)
Volume 96 **Industrial Power Systems with Distributed and Embedded Generation** R. Belu
Volume 97 **Synchronized Phasor Measurements for Smart Grids** M.J.B. Reddy and D.K. Mohanta (Editors)
Volume 98 **Large Scale Grid Integration of Renewable Energy Sources** A. Moreno-Munoz (Editor)
Volume 100 **Modeling and Dynamic Behaviour of Hydropower Plants** N. Kishor and J. Fraile-Ardanuy (Editors)
Volume 101 **Methane and Hydrogen for Energy Storage** R. Carriveau and D.S.-K. Ting
Volume 104 **Power Transformer Condition Monitoring and Diagnosis** A. Abu-Siada (Editor)
Volume 106 **Surface Passivation of Industrial Crystalline Silicon Solar Cells** J. John (Editor)
Volume 107 **Bifacial Photovoltaics: Technology, Applications and Economics** J. Libal and R. Kopecek (Editors)
Volume 108 **Fault Diagnosis of Induction Motors** J. Faiz, V. Ghorbanian and G. Joksimović
Volume 110 **High Voltage Power Network Construction** K. Harker

Volume 111 **Energy Storage at Different Voltage Levels: Technology, Integration, and Market Aspects** A.F. Zobaa, P.F. Ribeiro, S.H.A. Aleem and S.N. Afifi (Editors)
Volume 112 **Wireless Power Transfer: Theory, Technology and Application** N. Shinohara
Volume 114 **Lightning-Induced Effects in Electrical and Telecommunication Systems** Y. Baba and V.A. Rakov
Volume 115 **DC Distribution Systems and Microgrids** T. Dragičević, F. Blaabjerg and P. Wheeler
Volume 116 **Modelling and Simulation of HVDC Transmission** M. Han (Editor)
Volume 117 **Structural Control and Fault Detection of Wind Turbine Systems** H.R. Karimi
Volume 119 **Thermal Power Plant Control and Instrumentation: The Control of Boilers and HRSGs, 2nd Edition** D. Lindsley, J. Grist and D. Parker
Volume 120 **Fault Diagnosis for Robust Inverter Power Drives** A. Ginart (Editor)
Volume 121 **Monitoring and Control Using Synchrophasors in Power Systems with Renewables** I. Kamwa and C. Lu (Editors)
Volume 123 **Power Systems Electromagnetic Transients Simulation, 2nd Edition** N. Watson and J. Arrillaga
Volume 124 **Power Market Transformation** B. Murray
Volume 125 **Wind Energy Modeling and Simulation, Volume 1: Atmosphere and Plant** P. Veers (Editor)
Volume 126 **Diagnosis and Fault Tolerance of Electrical Machines, Power Electronics and Drives** A.J.M. Cardoso
Volume 128 **Characterization of Wide Bandgap Power Semiconductor Devices** F. Wang, Z. Zhang and E.A. Jones
Volume 129 **Renewable Energy from the Oceans: From Wave, Tidal and Gradient Systems to Offshore Wind and Solar** D. Coiro and T. Sant (Editors)
Volume 130 **Wind and Solar Based Energy Systems for Communities** R. Carriveau and D.S.-K. Ting (Editors)
Volume 131 **Metaheuristic Optimization in Power Engineering** J. Radosavljević
Volume 132 **Power Line Communication Systems for Smart Grids** I.R.S. Casella and A. Anpalagan
Volume 139 **Variability, Scalability and Stability of Microgrids** S.M. Muyeen, S.M. Islam and F. Blaabjerg (Editors)
Volume 145 **Condition Monitoring of Rotating Electrical Machines** P. Tavner, L. Ran and C. Crabtree
Volume 146 **Energy Storage for Power Systems, 3rd Edition** A.G. Ter-Gazarian
Volume 147 **Distribution Systems Analysis and Automation, 2nd Edition** J. Gers
Volume 152 **Power Electronic Devices: Applications, Failure Mechanisms and Reliability** F Iannuzzo (Editor)
Volume 153 **Signal Processing for Fault Detection and Diagnosis in Electric Machines and Systems** M. Benbouzid (Editor)
Volume 155 **Energy Generation and Efficiency Technologies for Green Residential Buildings** D. Ting and R. Carriveau (Editors)
Volume 157 **Electrical Steels, 2 Volumes** A. Moses, K. Jenkins, P. Anderson and H. Stanbury
Volume 158 **Advanced Dielectric Materials for Electrostatic Capacitors** Q. Li (Editor)
Volume 159 **Transforming the Grid Towards Fully Renewable Energy** O. Probst, S. Castellanos and R. Palacios (Editors)
Volume 160 **Microgrids for Rural Areas: Research and Case Studies** R.K. Chauhan, K. Chauhan and S.N. Singh (Editors)
Volume 161 **Artificial Intelligence for Smarter Power Systems: Fuzzy Logic and Neural Networks** M.G. Simões (Editor)
Volume 166 **Advanced Characterization of Thin Film Solar Cells** N. Haegel and M. Al-Jassim (Editors)
Volume 167 **Power Grids with Renewable Energy: Storage, Integration and Digitalization** A.S. Sallam and O.P. Malik
Volume 172 **Lighting Interaction with Power Systems, 2 Volumes** A. Piantini (Editor)
Volume 193 **Overhead Electric Power Lines: Theory and practice** S. Chattopadhyay and A. Das
Volume 905 **Power System Protection, 4 Volumes**

Artificial Intelligence for Smarter Power Systems

Fuzzy logic and neural networks

Marcelo Godoy Simões

The Institution of Engineering and Technology

Published by The Institution of Engineering and Technology, London, United Kingdom

The Institution of Engineering and Technology is registered as a Charity in England & Wales (no. 211014) and Scotland (no. SC038698).

First published 2021

The Institution of Engineering and Technology
Michael Faraday House
Six Hills Way, Stevenage
Herts, SG1 2AY, United Kingdom

www.theiet.org

British Library Cataloguing in Publication Data
A catalogue record for this product is available from the British Library

ISBN 978-1-83953-000-5 (hardback)
ISBN 978-1-83953-001-2 (PDF)

Typeset in India by MPS Limited

I dedicate this book to my wife, Deborah Doin, and to my children: Ahriel Godoy, Lira Godoy, Rafael Doin, and Luiz Notari. There is light and happiness in my life, because I have you.

Contents

About the author

Marcelo Godoy Simões is a Professor in Electrical Power Engineering, in Smart and Flexible Power Systems, at the University of Vaasa (Finland), in the School of Technology and Innovations, with the Electrical Engineering Department. He received a National Science Foundation (USA) CAREER Award, a very prestigious award for new faculty members in 2002. He was an US Fulbright Fellow at Aalborg University (Denmark) and worked as a visiting professor in several international institutions. He is an IEEE Fellow, published hundreds of journal papers and conference articles, and authored 12 books. He has pioneered the application of neural networks and fuzzy logic in renewable energy systems, his credential and publications are authority and relevant for advanced wind turbine control, photovoltaics, fuel cells modelling, smart-grid management and power electronics enabled power systems control for integration of renewable energy sources.

Foreword

It is my pleasure and privilege to write a Foreword for this important book on artificial intelligence (AI) techniques, and their applications in power electronics and power systems. It is well known that the AI techniques, particularly fuzzy logic and neural networks, have already established tremendous importance in power electronics and power systems, among many other industrial and nonindustrial applications. Particularly, their applications are very promising in the emerging next generation of smart grid and renewable energy systems. Again, among all the AI disciplines, it is expected that neural networks will have a maximum impact on power electronics and power systems. The area of power electronics, particularly, is very complex and multidisciplinary. The advancing frontier of power electronics with the AI technology will be challenging to the power electronics engineers. The book is authored by Prof. Marcelo Simoes, who is a world-renowned scientist in AI area. I am proud to mention that Dr. Simoes initiated his pioneering AI research in my power electronics laboratory in the University of Tennessee. It is interesting to note that in 1997, I organized a panel discussion session on advances and trends of power electronics in the IEEE Industrial Electronics Society Conference (IECON-1997), where I invited him as a panelist on AI applications in power electronics. He was the youngest panelist in such most important area. In the last 25 years, since his doctorate degree in 1995, he has established himself as a very prominent scientist in this area. The present book authored by Marcelo is very comprehensive. It extensively reviews the state-of-the-art technologies of fuzzy logic and neural network and their applications in power electronics and power systems. In addition, it includes real-time modeling and simulation, hardware-in-loop testing, deep machine learning, etc., which will be important in emerging smart grid and renewable energy systems applications. Of course, one of the nine chapters has been contributed by OPAL-RT engineers, who are specialized in this area. The book will be important for university professors and other professionals, and students who are doing research in this area. Of course, selected portions of the book can be taught in undergraduate and graduate programs. I wish success for this book.

Dr. Bimal K. Bose, *IEEE Life Fellow*
Emeritus Chair Professor in Electrical Engineering
(Formerly Condra Chair of Excellence in Power Electronics)
Member, US National Academy of Engineering
Department of Electrical Engineering and Computer Science
The University of Tennessee, Knoxville

Preface

I started this book many years ago, and it has been paused on and off due to so many other professional priorities, personal matters and evolving of my life as a whole. When I just thought that neural networks were saturated in power electronics and power systems, I observed the rapid evolution of deep learning, at the same time the maturity of smart-grid systems as a core in electrical engineering. I am very proud to introduce this book to our professional community. I hope all who read it, or have any brief consultation on any of the topics, will appreciate a solid foundation of artificial intelligence (AI), fuzzy logic, neural networks, and deep-learning for advancing power electronics, power systems, enhancing the integration of renewable energy sources in a smart-grid system.

When I graduated from Poli/USP in 1985 in Electrical Engineering, my expertise was electronic systems, high frequency circuits, and I was just starting to learn the basics of power electronics. Computer simulation was still based on mainframes, electrical circuit simulation in Spice, software was written in compiled languages, such as Pascal, C, FORTRAN. Designing and implementing a switching power supply required me to understand analog circuits of TVs, reading application notes of semiconductor companies, reverse engineering circuits from computers, taking notes on a notebook to document the design, and eventually burning and destroying many transistors and diodes during the workbench prototyping. I first learned to use MATLAB® in an IBM PC AT in 1988, and when I joined University of Tennessee for my Ph.D. program, I witnessed an evolution in how computer-simulation-based design and digital signal processing (DSP)-based hardware would enhance very complex control algorithms in real-life applications. From 1991 to 1995, I started to study, learn, and to apply fuzzy logic and neural networks in power electronics, enhancing wind energy systems, PV solar systems, and power quality diagnosis and energy management.

In my career, I have been writing books and publishing several papers; I saw how power electronics evolved and became a key enabling technology for the twenty-first century with the technology of smart-grids for the integration of renewable energy resources. The revolution in power electronics was introduced with solid-state power semiconductor devices in the 1950s. AI, initially on the first generation of neural networks, started about the same time, a few years earlier. During the 1960s, fuzzy logic was introduced by Lotfi Zadeh. With the emergence of microprocessors and later DSP controllers, there was a widespread application of power electronics in industrial, commercial, residential, transportation, aerospace, military, and utility systems. From the 1990s to now, we have had the age of industry automation, high efficiency energy systems that include modern renewable energy systems, integration of transmission, and distribution with bulk energy storage, electric and hybrid vehicles, and energy efficiency improvement of electrical equipment.

With the popularization of the backpropagation algorithm in 1985, a second wave of neural network research was made possible with so many topologies and architectures of neural networks, also many expert system shells, fuzzy logic systems for microcontrollers, and PLCS, eventually making the use of AI in power electronics and power systems a reality.

Power electronics is the most important technology in the twenty-first century, and our power systems, utility integration, and distribution systems became a power-electronics-enabled power system, with added intelligence to be a smart-grid system. In such a vision of smart grid, the role of power electronics in high-voltage DC systems, static VAR compensators, flexible AC transmission systems, fuel cell energy conversion systems, uninterruptible power systems, besides the renewable energy and bulk energy storage systems, has tremendous opportunities. In the current trend of our energy scenario, the renewable energy segment is continuously growing, and our dream of 100% renewables in the long run (with the complete demise of fossil and nuclear energy) is genuine. Therefore, the social impact of power electronics in our modern society is undeniable, and this book contributes with nine specialized chapters. After a general introduction in Chapter 1, there is a discussion in Chapter 2 of how hardware-in-the-loop, real-time simulation, and digital twins are enabling future smart-grid applications, with a strong need for AI. Chapters 3, 4, and 5 present everything necessary for an engineer to develop, implement, and deploy fuzzy systems, with all sorts of engine implementations, and how to design fuzzy logic control systems. Chapters 6 and 7 focus on feedforward neural networks and feedback, competitive and associative neural networks, with methods, procedures, and equations, discussed in an agnostic and scientific perspective, so the reader can adopt and adapt the discussions into any modern computer language. Chapter 8 discusses the applications of fuzzy logic and neural networks in power electronics and power systems.

During the twentieth century, particularly after the advent of computers and advances in mathematical control theory, many attempts were made for augmenting the intelligence of computer software with further capabilities of logic, models of uncertainty, and adaptive learning algorithms that made possible the initial developments in neural networks in the 1950s. However, a very radical and fruitful of such foundations was initiated by Lotfi Zadeh in 1965 with the publication of his paper "Fuzzy Sets." In such paper, the idea of membership function with a foundation on such a multivalued logic, properties, and calculus became a solid theory and technology that bundled together thinking, vagueness, and imprecision. Every design starts from the process of thinking, i.e., a mental creation, and people will use their own linguistic formulation, with their own analysis and logical statements about their ideas. Then, vagueness and imprecision are considered here as empirical phenomena. Scientists and engineers try to remove most of the vagueness and imprecision of the world by making clear mathematical formulation of laws of physics, chemistry, and the nature in general. Sometimes it is possible to have precise mathematical models, with strong constraints on non-idealities, parameter variation, and nonlinear behavior. However, if the system becomes complex, the lack of ability to measure or to evaluate features, has a lack of definition of precise modeling, in addition to many other uncertainties and incorporation of human expertise, making almost impossible to explore such a very precise model for a complex real-life system. Fuzzy logic and neural network became the foundation for the newly advanced twenty-first century of smart control, smart modeling, intelligent behavior, and AI. This book presents the basics and foundation for fuzzy logic and neural network, with some applications in the area of energy systems, power electronics, power systems, and power quality.

Fuzzy control has a lot of advantages when used for optimization of alternative and renewable energy systems. The parametric fuzzy algorithm is inherently adaptive, because the coefficients can be altered for system tuning. Thus, a real-time adaptive implementation of the parametric approach is feasible by dynamically changing the linear coefficients by means of a recursive least-square algorithm repeatedly on a recurrent basis. Adaptive versions of the rule-based approach, changing the rule weights (Degree of Support) or the membership functions recurrently is possible. The disadvantage of the parametric fuzzy approach is the loss of the linguistic formulation of output consequents, sometimes important for industrial plant/process control environment.

Fuzzy and neuro-fuzzy techniques became efficient tools in modeling and control applications. There are several benefits in optimizing cost effectiveness because fuzzy logic is a methodology for handling inexact, imprecise, qualitative, fuzzy, verbal information such as temperature, wind speed, humidity, and pressure in a systematic and rigorous way. A neuro-fuzzy controller generates, or tunes, the rules or membership functions of a fuzzy controller with an artificial neural network approach.

For applications of alternative and renewable energy systems, it is very important to use AI techniques because the installation costs are high, the availability of the alternative power is by its nature intermittent, and the system must be supplemented by additional sources to supply the demand curve. There are efficiency constraints, and it becomes important to optimize the efficiency of the electric power transfer, even for relatively small incremental gains, in order to amortize installation costs within the shortest possible time. Smart-grid systems must be evaluated in comprehensive case studies, engineering analysis, big databases, with detailed modeling, and simulation techniques.

In this third decade of the twenty-first century, we want young students and junior engineers to become motivated by the third-wave of research in neural networks, i.e., big data analytics, data science, and deep learning. Chapter 9 is extensive in discussing deep learning and big data applications in electrical power systems. The approach is comprehensive, clear, allowing implementation in any hardware and software. The reader will learn what is a deep-learning neural network, how it can be used for classification, regression, clustering, and modeling. How convolutional neural networks can be used for smart-grid applications, and how the previous paradigm of recurrent neural networks has been modernized in the twenty-first century with long short-term memory neural networks (LSTM) and how fuzzy parametric CMAC neural networks can also be applied for current deep-learning AI revolution.

All the chapters review the state of the art, presenting advanced material and application examples. The reader will become familiarized with AI, fuzzy logic, neural networks, and deep learning in a very coherent and clear presentation. I want to convey my sincere enthusiasm with this hopeful timeliness book in your hands. I am very confident that this book fulfills the curiosity and eagerness for knowledge in AI for making power systems, power electronics, renewable energy systems, and smart grid, a legacy for generations to come in this century.

I am grateful to all my past undergraduate and graduate students, most of them are currently working in high technology and advanced in their careers; we became colleagues and professional fellows. I am grateful to all faculty and researchers who have been working with me in this professional journey in the past a little more than three decades in my life. There are so many of you, important in my life, that it is not fair to list names, but we know each other and we support each other.

Specifically, I am very thankful to the support of Dr. Tiago D.C. Busarello who reviewed the manuscript and gave me suggestions for improvements. I am grateful to Alexandre Mafra who kept his professional dream in working with neural networks and gave me valuable feedback. To the group of colleagues and engineers in OPAL-RT and the guest authors of Chapter 2, I show my strong appreciation and gratitude for the collaboration, I am especially grateful to Prof. Bimal K. Bose, my former Ph.D. adviser, who motivated me a few years ago to write this book.

I am grateful, *in memoriam*, to Dr. Paulo E.M. Almeida; he was my Ph.D. student, he became a successful professor and a leader in intelligent automation.

I dedicate this book to you, reader, such a knowledge is for you to advance, for you to make our world better, for you to make our society more prosperous. Thank you for reading this book.

—Marcelo Godoy Simões, Ph.D., IEEE Fellow,
Professor in Electrical Power Engineering, in Smart and Flexible Power Systems, at the University of Vaasa (Finland), in the School of Technology and Innovations, with the Electrical Engineering Department.

Chapter 1
Introduction

1.1 Renewable-energy-based generation is shaping the future of power systems

Over the last two centuries, the modern society has consumed more energy per capita than all the previous cultures throughout history. Such an industry revolution brought an unprecedented standard of living, dependent on fossil-fuel deposits formed millions of years ago. The control of fossil-fuel energy reserves has been the central policy of governments and geopolitics for the past twentieth and now in the twenty-first centuries. The industrialized countries and international companies have been enjoying such growing wealth. In contrast, there are a majority of people in the world who have been denied such access, and at the same time our modern society has seen depth into poverty, increased marginalization, associated with conflicts, wars, that made people displaced from their original places of birth. For the rich and emerging countries, renewable-energy-based power generation is already shaping their future. For the globalized world, such a new paradigm of distributed power generation also has the potential to make them increasing standards of living, with grassroots' local sustainable growth. The industrial revolution of the nineteenth century was sequel by the technological revolution of the twentieth century, and it is our compromise now to have the energy revolution of the twenty-first century, making access to electric power with resources provided by the nature, sustainable and renewable, such as solar, wind, hydro, geothermal, distributed with inclusion of all people, to develop energy solutions that are sustainable, and socially fair.

Everything defining our way of life requires an energy-consuming apparatus made with materials that have been transformed from fauna, flora, and earthly resources. There is an energy chain behind all transformations from earth resources into manufactured products and goods. The food we eat, the clothes we wear, the buildings we live, the appliances we use, cars and everything necessary for the modern society require energy to make them or require power to operate them. Therefore, a standard indicator of economic production would be the ratio of energy consumption by the people as final user. It is assumed that such per capita energy consumed is nearly a constant number. Developed countries, when compared to emerging economies, even with 200-fold difference in their income per capita, will make the ratio of income to energy consumption to be from 0.30 to 0.32, supporting that increasing standards of living also increase energy needs, making such ratio constant.

1.2 Power electronics and artificial intelligence (AI) allow smarter power systems

The technology of power-electronics-enabled power systems, called a microgrid, or a smart grid, is fundamental for renewable energy systems because the intermittent nature of renewable energy affects the output characteristics of a generator and converter sets (i.e., their voltage, frequency, and power) and must be compensated by the integration with energy storage. Energy storage is always needed during a surplus of input power and compensation in the case of a lack of input power.

Operators of conventional transmission systems based on hydro- and thermal plants can easily dispatch optimal generator settings to meet daily load variations and scheduled maintenance. This obviously is not possible for power grids with a large proportion of nonscheduled power sources such as solar- and wind plants. Consequently, large storage units must compensate for their intermittency. Depending on the available sources, inverters, rectifiers, and DC/DC converters are required. A rectifier might be a front end for an electric grid connected to a load, or an inverter can be the interface with the local generation. There are other converters for intermediate stages necessary for adapting the energy produced by the source in such a way that both the energy source and the inverter operate at their highest efficiency.

Power-electronics-enabled power systems allow smarter power systems, and artificial intelligence (AI) supports high performance and reliability, highly efficient aligned with developing technologies, encompassing water, food, bio-resources, renewable energy, waste, where smart-grid systems will advance ecosystems to human services, through a circular economy paradigm.

Wind energy production grew exponentially over the past few years. In the United States, the rated installed wind power has more than doubled over the past decade. The first utility-scale wind turbine (WT) was a 50-kW WT installed in the early 1980s, since then WT manufacturers have been pushing the boundary of the maximum output power that can be produced by a single WT. Currently, by the year 2020, WTs with rated output power as high as 10 MW are available in the market. Moreover, WTs in the 15–20 MW range are under development and expected to be available for deployment.

While the electric generators used in multi-MW WTs can have rated output voltages as low as 690 V, their rated powers are as high as 10 MW; in these cases, conventional power electronics interfaces such as back-to-back voltage source converters (BTB-VSC) must be designed to withstand currents in the kA range, per-phase. The state-of-the-art semiconductor technology cannot produce single switches that withstand current in such range; instead, multiple switches are connected in parallel to split the current evenly. However, connecting switches in parallel may not guarantee the dynamic sharing of current among the switches. A standard solution is to combine multiple BTB-VSCs in parallel, eliminating the need to connect individual switches in parallel, increasing the cost and weight of the WT, and reducing the overall efficiency, in addition, BTB-VSC requires filters.

Future applications of power electronics in the medium-voltage integration of renewables and energy storage systems require that converters should reduce their rated components, improve efficiency, and reliability with further integration of renewable sources. The nominal voltage value in medium-voltage applications for regular distribution power systems is on the order of 4 kV, with possible 400-V-rated battery systems to be integrated by using specialized power converters, typically using 650-V MOSFETs in their primary side, with 3–5 kV IGBTs on their secondary side. The improvement of efficiency and reliability is significant to make converters to operate in a bidirectional manner, connected to the utility grid, with possible integration of local consumers capable of generating local power (prosumers), and intelligence to manage, control, and interconnect with the utility and centers of electrical energy distribution. In this regard, it is essential for students, engineers, researchers, industry, and commerce stakeholders, to understand the foundation of fuzzy logic and neural networks for enhanced power systems.

There is an evolution of the power system, from unidirectional power flow, current technology, and future bidirectional power with integrated communications and advanced infrastructure. The power grid is evolving by shifting energy supply from large central generating stations to smaller distributed energy resources. Therefore, power electronics is the enabling technology to allow power systems based power electronics for smart-grids.

1.3 Power electronic, artificial intelligence (AI), and simulations will enable optimal operation of renewable energy systems

Integrating and operating several bidirectional power electronic systems in large grids is an engineering challenge. The performance of individual energy conversion system and the performance of the global grid integrating several power electronic converters must be evaluated before they are built. Modeling and simulation using numerical or analog methods must then be used to predict the performance and reliability of the contemplated systems under several normal and faulty operating conditions. The results of these simulations will be used to specify each component in building, commissioning and operating in the real-life system.

AI-based control system is considered as a tool to help operating such complex systems to ensure optimal and secure operation, i.e., optimize the operating not only to achieve the best economic performance but also, which is more challenging, to minimize the risk of cascaded failures following fault conditions. Implementing AI-based control to maintain system security will, however, require accurate models of the global systems to rapidly test the performance of several contingencies and to predict the optimal control actions such as load and line switching, transformer tap changer and generator settings, and many other parameters. The model will also be required to train the AI fuzzy logic system since that actual grid system cannot be used to test several severe contingencies to train the system.

The AI system will also be necessary to validate the accuracy of system models and their variations over time due to failure, equipment aging and maintenance.

Power electronic enables the integration of renewable energies, AI-based controllers will enable the optimal operation of these complex systems, and fast simulation using accurate models will be mandatory to implement, design, and test AI-based system. Introduction to AI-based systems and real-time plus hardware-in-the-loop simulations are the main subjects of this book.

1.4 Engineering, modeling, simulation, and experimental models

Engineering is a body of knowledge where theoretical or an experimental model is supported for example by the balance equations for stored masses, energies, and impulses, or by physical–chemical constitutive equations. Modeling and analysis can be conducted using phenomenological equations of irreversible processes (thermal conductivity, diffusion, and chemical reaction), and applying entropy balance equations when irreversible processes are interrelated (very often in energy conversion). However, electrical circuit analysis and design are typically supported by the connection and state-equations associated with describing such an interconnection of process elements.

Designing power-electronics-enabled power systems to support smart-grid systems must also consider the integration of renewable energy sources to the system. Numerical analysis-based software is available for modeling energy systems and power electronics, for studying their dynamics and transient performance using state-space or nodal analysis, or for their linear algebraic systems when studying their static or steady-state solutions.

There are other advanced paradigms for the simulation of multiphysics domain in object-oriented programming, using software for differential algebraic systems aiming noncausal modeling with mathematical equations and the use of object-oriented constructs to facilitate reuse of modeling knowledge. The current state of the art in computer simulation allows a hybrid model to be performed with mathematical analysis containing differential algebraic equations (DAE), modified nodal analysis matrices, ordinary differential equations (ODE), with time responses of all the time-varying variables in the mathematical model. However, applying conventional techniques to simulate complex power electronic systems is time-consuming since system topology and therefore system state-space equations and resulting matrices change each time that switch status is changed. Since several switches are constantly tuning-on and -off at high frequencies, processing time can become excessive even for the simulation of a single converter system. The time to simulate an one-second phenomena can easily take hours for large power grids with multiple converters, as expected for wind- and solar plants integrated with distribution or transmission systems.

Testing of actual power electronic control systems can be done using a physical setup representing the actual system. Such analog simulation is limited to analyze performance of rather simple systems and to perform testing power electronic system controls integrated with large grids requires numerical simulation of the grids with the control systems. If actual control system hardware must be tested, then these actual control systems are connected to a numerical model of the plant implemented on a real-time simulator. This is called hardware-in-the-loop (HIL) simulation as explained in Chapter 2. Such simulation requires hard real-time constraints, i.e., a discrete time result must command at their real clock time, in order to maintain simulation accuracy, i.e., to make sure that the controllers under test will react like if it was connected to an actual power grid. Furthermore, using a grid simulator in conjunction with AI-based system will require that the simulation process reaches the speed faster-than-real-time to perform the maximum number of analysis in minimum time. Advanced simulation software taking the advantage of parallel processing on cloud infrastructure will be necessary.

1.5 Artificial intelligence will play a key role to control microgrid bidirectional power flow

AI is a broad and encompassing field, consisting of machine learning, logic-based tools used for knowledge representation and problem-solving, knowledge-based tools, based on databases, information, and rules, probabilistic methods: tools that allow agents to act in incomplete information scenarios, embodied intelligence: engineering toolbox, which assumes that a body (or at least a partial set of functions such as movement, perception, interaction, and visualization) is required for higher intelligence, search and optimization: tools that allow intelligent search with many possible solutions.

Fuzzy logic and neural networks have been at the core of technologies that solve problems as multi-agent systems (MAS), agent-based modeling, teaching computers how to learn a task that involves several techniques to deal with smart-grid applications. A universal Pareto principle emerges to support that 80% (even more) of current efforts and results are driven by 20% of the technologies. The approach taken in this book in understanding and applying real-time modeling and simulation of power-electronics-enabled power systems, fuzzy logic, neural networks in feedforward and feedback, plus using deep learning will allow a vast spectrum of knowledge to design smarter power systems.

Control strategies for microgrids and smart grids will allow power flow bidirectionally into the grid, regulate the power sharing among distributed generators, and comply with power quality requirements. The power flow control must be accomplished at every instant and in both islanded and grid-connected modes. For this reason, the operating performance of any distributed intelligence of smart grids strongly relies on the employed control strategy. The original power-based control is currently based on lead-follower architectures and will soon be distributed on the basis of multi-agent systems (MAS). For the current technology, a power processor placed at the point of common coupling (PCC) of the microgrids will allow a central converter to interact with an energy storage station.

There are many required measurements at each side of the PCC, data packet information sent by each active node, output quantities and its maximum generation available, and converter rate capacity. The grid side reference to dispatch the microgrid is defined and set by the highest levels of energy management control in the microgrid. The intelligence should also allow smart metering, i.e., the converter between the local source/load and the grid to be capable of tracking the energy consumed by load or maybe the amount of energy injected in the grid. Real-time information must be passed to an automatic billing system capable of considering parameters as the buy/sell energy in real time at the best economic conditions and informing the owner of the installation of all required pricing parameter decisions. Communication is necessary for the intelligent functioning of smart power systems, depending on their capability to support communications at the same time that power flows in the systems. Such functions are fundamental for overall system optimization and for implementing sophisticated dispatching strategies. Fault tolerance is important in avoiding the propagation of failures among the nodes and to recover from local failures. This capability should be managed by the power converter, which should incorporate monitoring, communication and reconfiguration systems, and extra intelligent functions capable of making the user interface friendly and accessible anywhere through Internet-based communications.

Designing a smart grid, a power-electronics-enabled power system, enhanced control with artificial intelligence requires understanding the hierarchical energy system with real-time constraints, therefore, the reader is encouraged to study first Chapter 2 of this book. OPAL-RT is very powerful software, with a comprehensive suite of solutions, including parallel simulation technologies, capable of simulating very large grids and power electronic systems faster-than-real-time. MATLAB®/ Simulink® with several toolboxes is fundamental in allowing circuits, block diagrams in a solid electrical engineering perspective. MATLAB/Simulink can be integrated with fuzzy logic and neural network toolboxes, and recently MathWorks developed deep learning algorithms. Machine learning has typically other computational environments, and Python is very often utilized with so many scientific libraries. Particularly, for a deep learning framework there are currently two main streams, TensorFlow (Google) and PyTorch (Facebook), both of which are exceptional solutions. Within the TensorFlow approach, there is a library called Keras for deep neural networks.

1.6 Book organization optimized for problem-based learning strategies

This book can be taught in the natural sequence, i.e., from Chapter 1, sequentially toward Chapter 9, giving a comprehensive scientific perspective of power electronics, power systems simulation, digital twins, fuzzy logic, neural networks, and deep learning. It can be used in a course based on problem-based learning strategies, where the instructor could develop a few complex lectures and projects based on project-based learning methodologies. All the chapters have a brief introduction on the theoretical background, a description of the problems to be solved, objectives to be achieved, the visualization of block diagrams, electrical circuits, mathematical analysis or computer code that approaches a high-level discussion, which can be complemented with further theoretical perspectives in electrical engineering as well as in computer science.

If the instructor or reader wants to focus this book on fuzzy logic, the recommended sequence of chapters is 1, 2, 3, 4, 5, and 8. On the other hand, a focus on neural networks may have a sequence of chapters such as 1, 2, 6, 7, 8, and also 9 for deep learning as there is a strong potential in using deep learning for smarter power systems, which can be studied independently from the other chapters (if the reader has already background in Neural Networks). This book has nine chapters as follows:

- Chapter 2 covers the use of real-time simulation for smart-grid applications utilizing power-electronics-enabled power systems. It introduces the concepts and applications related to real-time simulation and HIL testing and discusses some specific use cases. The use of fast or real-time simulation in advanced applications such as the generation of datasets through batch testing to train fuzzy logic or neural-network-based control and the perspective of digital twins to improve future smart-grid system resilience.
- Chapter 3 discusses intelligent systems, fuzzy reasoning, introduction to fuzzy sets and fuzzy logic, and fuzzy logic kernel operations.
- Chapter 4 presents fuzzy modeling with fuzzy reasoning, rule-based and relational-based, fuzzification, defuzzification with the fuzzy inference engine, Mandani or Type 1 Fuzzy technique, parametric or Takagi–Sugeno–Kang technique, fuzzy operations in different universes of discourse, and extensions on neural-fuzzy approaches.
- Chapter 5 provides implementation methodologies of fuzzy-logic-based control, discussion of heuristic identification of controllers, implementation of fuzzy PI and fuzzy PD, industrial fuzzy control supervision and scheduling, and hybrid parametric fuzzy PID control.
- Chapter 6 introduces feedforward neural networks, with a techno-historical perspective, description of the backpropagation algorithm, examples of binary classifier and mapping using McCulloch–Pitts multilayer perceptrons, discussions on error measurement for convergence training, several possible neuron activation functions, data processing for neural networks, and neural-network-based computing.
- Chapter 7 shows feedback, competitive, and associative neural networks, discussions on linear vector quantization networks, self-organizing maps, Kohonen networks, counter-propagation networks, probabilistic neural networks, and industrial applicability of artificial neural networks.
- Chapter 8 details applications of fuzzy logic and neural networks in power electronics and power systems, discussing the three main industrial fields of fuzzy logic and neural network for controller design, function optimization, and function approximation, with advanced adaptive neural fuzzy inference

systems, and discussions on AI-based control systems for smarter power systems describing how after neural network system identification is performed, a control system can be implemented with three possible architectures, model predictive control, adaptive inverse-model-based control, and model reference control.

- Chapter 9 introduces deep learning and big data applications in electrical power systems, discussing big data analytics, data science, engineering and power quality, big data for smart-grid control, online monitoring of diverse time-scale fault events for non-intentional islanding, smart electrical power systems with in-depth learning features, how to use deep learning for classification, regression, and clustering, details on the implementation of convolutional neural networks (for deep multidimensional algebraic mapping), and how recurrent neural networks have been transformed from earlier back-propagation-through-time to modern long short-term-memory-based recurrent neural networks for deep recurrence of high order systems, or based on text recurrence, or streaming data or audio–video on industrial high-speed data applications; Chapter 9 concludes discussing computer-based versus cloud-based implementation, computation with discussions on fuzzy parametric CMAC neural network for deep learning, allowing computer-capable graphics processing units hardware for multiprocessing of very complex dynamical systems.

We expect this book to be a reference for anyone interested in understanding, designing, or analyzing technology of fuzzy logic, neural networks, deep learning, in power systems and power electronics. It also presents an introduction to real-time-systems (RTS) and HIL modeling analysis for renewable and distributed energy systems. RTS could be used to test the concept and performance of AI and fuzzy-logic-based control systems before their implementation in the field. This book supports the analysis model and design for the new generation of smarter power systems and smart-grid technology. This work can be adopted as a textbook for an advanced undergraduate course or master level graduate course, the instructor may develop exercises to complement the educational use of this book.

Chapter 2

Real-time simulation applications for future power systems and smart grids

2.1 The state of the art and the future of real-time simulation

Real-time simulation has been used in various industries for decades now, but it is particularly used in electrical transmission systems. These dynamic simulation tools can be divided into two main categories: (i) transient stability (TS) tools based on fundamental frequency phasors and (ii) electromagnetic transient (EMT) tools allowing simulation of fast transient and higher frequency phenomena.

2.1.1 Transient stability tools for off-line or near real-time analysis

Off-line TS tools are used to assess the capability of the system to survive to the next event, considering the present state of the power grid and other contingencies. Most of these tools are based on fundamental frequency positive-sequence phasors only, with simplified models. They can simulate the dynamics of the system considering conventional generator rotation speed variations (electromechanical transients) and the effect of voltage and speed regulators as well as the effect of large perturbations.

These tools are used by planning departments to design power systems, and by system operators in control rooms as parts of sophisticated dynamic security assessment (DSA) tools to estimate the maximum power transfer capability (MPTC) of the system at each instant. They are often referred to as real-time assessment tools (although they are technically near real time), since they are interfaced with the supervised control and data acquisition (SCADA) and state estimators to compute the MPTC every 5–10 min. Various technologies are used to accelerate the computation to estimate the MPTC for the maximum number of contingencies within the 5–10 min update period. Parallel processors can be used to compute the stability of various contingencies by executing simulations in parallel on large servers.

For the simulation of very large power systems, large computational times are often required to achieve the simulation of all contingencies. The precalculation of thousands of contingencies may be required in certain cases, for instance, to implement rule-based automation algorithms. TS tools are currently considered mature and are widely used in the power systems industry.

2.1.2 Transient stability simulation tools for real-time simulation and HIL testing

It is important to test actual or planned control and protection equipment by interfacing it to a simulated power grid. In such a case, the simulated power grid must react in a "hard real-time" fashion to test the dynamic response of the closed-loop system formed by the controller and the grid. This is called hardware-in-the-loop (HIL) testing and is further discussed in this chapter. Real-time TS tools must thus be capable of using parallel processing to achieve hard real-time calculations within time steps to the order of 1 ms. Very few real-time TS tools are available on the market. One of them is ePHASORSIM by OPAL-RT (Jalili-Marandi *et al.*, 2013).

2.1.3 Electromagnetic transient simulation (EMT) tools—off-line applications

EMT tools are required to design high-voltage lines and to study the integration of FACTS and HVDC transmission interconnections. Both distribution and transmission systems require the study of the integration of inverter-based renewable energy sources, such as solar and wind. The power systems industry has also adopted EMT with varying levels of detail, allowing analysis of a range of phenomena, from the slowest to the fastest. Conversely to TS tools that can efficiently simulate electromechanical phenomena with oscillation frequencies in ranges below the fundamental frequency (60 and 50 Hz), EMT analysis allows the simulation of oscillations up to ranges of 1 kHz and above, which includes switching transients and dynamics of power electronic systems.

Most of the EMT simulation tools are used in the planning departments of electrical power utilities and by equipment manufacturers to determine stresses on all components and to design and test control and protection systems of conventional and power electronic equipment. Since power grids are becoming larger and are integrating more power electronic equipment, it is essential to improve these off-line tools to maximize their computation capability. Research on advanced solvers is ongoing, and some off-line EMT simulation software manufacturers are contemplating the implementation of parallel processing to accelerate their simulations.

2.1.4 Electromagnetic transient simulation (EMT) tools—real-time HIL applications

Real-time EMT simulators were once (some may still be) fully analog and were designated as transient network analyzers, as illustrated in Figure 2.1. The term "real-time simulator" (hereafter RTS) now refers mostly to digital simulators and signifies that complex systems may be simulated in real time using powerful parallel computers interfaced with real-world equipment. In general terms, testing real controller and protection system equipment using a simulated plant is referred to as HIL testing. Real-time simulation has become a key tool in power systems with the emergence of the integration of FACTS and HVDC technologies in transmission

Figure 2.1 Transient network analyzer, Hydro-Québec Research Institute, 1976

systems (Li *et al.*, 2015; Zhu *et al.*, 2014; Vernay *et al.*, 2017), in which the precision timing of power electronic devices and flexible power flow controls requires that power system stability to be addressed and remedied.

Several utilities such as Hydro-Québec (HQ), RTE (the French TSO), CEPRI (China), and Entergy (USA) have implemented large real-time simulation laboratories equipped with actual replicas of HVDC and FACTS control systems interconnected with RTS to perform HIL tests. Such laboratories are becoming necessary to verify the dynamic performance of complex systems integrating several HVDC and FACTS controllers supplied by manufacturers. RTSs using control replicas are necessary to analyze phenomena such as HVDC inverter commutation failures following faults, and essential to validate the proper interaction of the equipment controls and protection with the rest of the system. RTS with control replicas are also used for maintenance, when it is necessary to verify the impact of controller modifications before implementing them in the field as well as to explain and solve control instabilities not detected during the design and commissioning phases. Finally, these laboratories allow for training of personnel responsible of advanced studies, testing, and field maintenance.

2.1.5 Shift in power system architecture with increased challenges in system performance assessment

Because of transmission line complex meshing and the long distances required for high voltages, transmission systems have always been complex and have driven innovation toward technologies that can, for instance, control power flow, increase

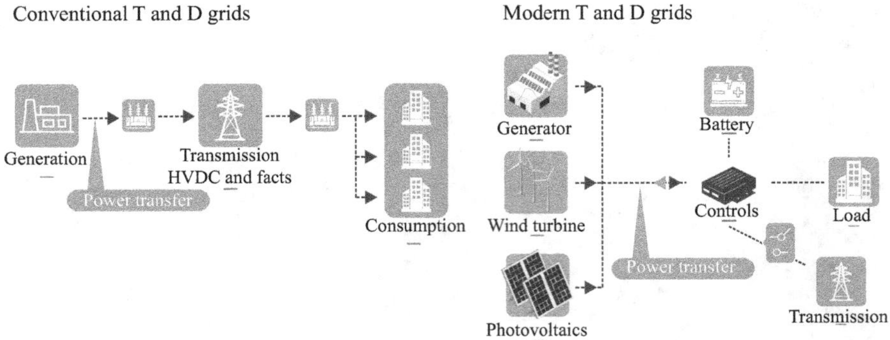

Figure 2.2 Comparing conventional and modern power grids

power transfer margins, allow flexible compensation, and regulate frequency and voltage. Such technologies include fast generator voltage regulators, HVDC transmission and interconnection as well as FACTS, fast local protection systems, and special wide area protection and control systems requiring complex and reliable communication systems.

Distribution systems, conversely, have conventionally been configured in radial configuration with short lines and simple controls and with power flowing only in one direction: from the substation to the client.

Transmission and distribution power systems are, however, experiencing an important shift away from conventional power system structures, toward modern grids centralized generation (using large rotating machines) which is now being complemented by distributed generation using power electronics. To achieve the objective of making the grid smarter, it is now the distribution and distributed generating systems that face the most complex challenges, as generators of all types and sizes with increasing amounts of power electronics are installed in a distributed fashion. Figure 2.2 illustrates some differences between conventional and modern electric power grids.

The addition of power electronics–based generators in transmission and distribution grids reduces the total inertia of the power systems and consequently decreases the response time following events, as the total kinetic energy stored in rotating machines relative to the total power capability is smaller. Consequently, one of the big challenges is to evaluate the global system performance and its capability to survive disturbances, as system response is much faster and highly dependent on power electronic control interactions.

2.1.6 EMT simulation to improve dynamic performance evaluation

It is now recognized that the assessment of power system stability with the increasing penetration of renewable energies requires simulation tools that go beyond traditional TS methodologies. In fact, studies now show that EMT-type tools are necessary and are used by various utilities (Wind Energy Systems Sub-Synchronous Oscillations: Events and Modeling, n.d.). Until direct methods are developed, the safest method to evaluate system dynamic security of low-inertia power grids is to use EMT models and simulation tools capable of simulating the details of fast power electronic systems. This is so, since fast power electronic control and protection systems react to EMTs, which may affect the overall dynamic performance of the grid and the power transfer capability evaluation. However, EMT simulation requires simulation time steps in the range of 10–100 μs, which leads to very large processing times when system size increases, unless parallel processing is used.

These new challenges drive an increased adoption of faster-than-real-time EMT simulation using parallel processing methods to implement HIL testing and real-time simulation technology for the design, analysis, and verification of smart-grid power equipment and controllers. As simulator computing power increases, more complex analysis can be performed at a lower cost; thus, this technology is now used or being contemplated by large utilities dealing with power grids integrating large quantities of power electronic systems to analyze the ability of the system to survive thousands of contingencies. Fast EMT RTS are also used in national research centers and universities worldwide for both researching and teaching advanced concepts in power systems. This prepares future engineers to contribute to the continuous improvement of power systems throughout their careers and opens the door for them to innovate extensively.

2.1.7 Fast EMT RTS as an enabler of AI-based control design for the smart grid

Artificial intelligence (AI)-based controls require a tremendous amount of data for training, although some may be acquired over time by monitoring the actual grid. However, simply monitoring the grid may not be sufficient to train AI on some critical cases—so-called edge or corner cases—the training coverage of which must be ensured. RTS technology can help in providing a sufficient volume of quality data to train AI as well as to test its implementation. One should also consider that AI-based controllers will be used locally close to equipment and remotely in control centers. Communication systems will therefore also play a critical role.

In short, modern power grids will become smarter with faster dynamic response but will be much more complex to control for ensuring specified service quality metrics that include the capability of surviving system faults. Consequently, AI-based control and protection systems will be used together with fast simulations and communication system technologies to increase the security of the power system ultimately.

AI is very promising as an improvement tool for smart-grid reliability and security, and it is important to learn about solutions that will help in deploying these technologies. This chapter introduces one of the most efficient tools to help in releasing better AI-based controls and protections to the smart grid. It presents some notions of real-time and accelerated (faster-than-real-time) simulation utilizing RTS technologies equipped with high-end multicore processors, field-programmable gate array (FPGA) chips, and cloud computing. Examples of the use of RTS for different applications, including smart inverters and wide area protection and control, are presented. The concept of digital twins (DTs), which enmeshes AI techniques in RTS technologies, is also discussed in the chapter.

2.2 Real-time simulation basics and technological considerations

2.2.1 Notions of real time and simulation constraints

One is tempted to ask the following question: "Time is already real; so, why do we say 'real time'?" In fact, time is a constant vector from past to present and from present to future. Time plays an important role in the cognition process, as one observes the sequence of events and measures the rate of change in movement, sounds, and images. This passage of time is what one experiences in a 120-min movie in which the characters witness 10 years of their lives elapse over its duration—an entertaining compression of time.

In power system analysis, an example of a manipulation of time would be post-fault analysis using waveforms quantifying variations in voltages, currents, frequencies, etc., analyzed in time-domain graphics, in periods spanning from cycles to minutes. In physics, real time is time as measured on a clock. It has a cadence; it is exact; it is "on time." In computing technologies, a real-time system is one for which the calculation results depend not only on its calculations and the logic involved, but also on the time within which these results are obtained. Real-time performance is also highly dependent on the timely reception of input and output signals exchanged between a real-time system and anything with which it interacts. To understand this in the context of real-time simulation, it may still be further divided into subcategories. Here, let us cover what is meant by the linked concepts of soft and hard real-time systems for mission-critical and noncritical applications.

In the case of "soft" real-time systems, an unsynchronized output from the computing device results in a degradation of the simulation quality, without necessarily resulting in critical errors. Soft real time is largely found in consumer technology applications. In video games, especially in action games, the notion of real time is very important, as it ensures an optimal player experience. The dynamics involved—from image processing by the human brain to motor response transferred to the gamepad—need to produce an immersive experience together, leaving it to the player to properly coordinate the sensory input with motor interactions and to win or lose as a consequence. Computing performance and latency seen in online games, for instance, may worsen the user experience, but the application is not mission-critical, and the user adapts to glitches. In gaming systems, the rate and resolution of images exchanged can purposely be degraded and/or controlled to enhance interactivity.

As a comparison, one could see the control room of a power system similarly, by substituting the gaming system itself with the power system and its ensemble of monitoring devices. Some control operations can be performed manually and do not require fast control that leaves it to the operator to process the information and act accordingly. However, excessive communication delays or failure of the monitoring system can largely degrade the quality of service and even cause unwanted downtime. So, the process involving power system operator manual controls can be categorized as a soft real-time application, although as compared to video games, it is mission-critical and must involve fail-safe equipment to ensure reliable continuity of service.

2.2.2 Concept of hard real time

Computing systems involving "Human-in-the-Loop" interactions, such as gaming systems or flight training simulators, may require calculation and information transmission times, for instance, of 15–30 ms or below (depending on the application) to match the average reaction time of humans (and to reproduce a realistic look and feel). This feedback delay time in human perception is, according to medical studies, somewhere between 150 and 300 ms (Jain *et al.*, 2015), depending on various factors. Similarly, for controller HIL (CHIL: as defined in Section 2.1.2) systems, the simulator calculation and communication times should be adapted to the fastest functions of the control system under test. Some control systems have time constants between 10 and 100 ms and others have time constants below 100 μs. In all cases, the RTS used to test controls must compute results fast enough for the controller to respond to realistic stimuli with realistic reaction times.

In systems with hard real-time constraints, each discrete time result must be published from the computing platform exactly at its corresponding wall-clock time, or errors may be there. Such failure to meet real time is called an overrun, which generally leads to a failure of the test run itself. In some cases, the overrun produces glitches in voltage and current waveforms, which may be confused with bad controller design. It is therefore important that the RTS provides overrun detection and that, most importantly, the simulation system is optimized for the desired application to produce reliable results.

The *sine qua non* condition for obtaining real-time simulation is the capability of the RTS to achieve hard real-time performance. For power grids, RTS must accurately simulate electromechanical and EMT phenomena expected in complex systems with electric and power electronic subsystems. A typical time step ranging from 10 to 100 μs or lower is required to ensure accurate results. Such performance is still a challenge and requires optimized parallel computer architectures as described next.

2.2.3 Real-time simulator architecture for HIL and its requirements

Real-time simulation of power systems can be defined as the reproduction of output (voltage/current) waveforms, with the desired accuracy, the representative of the behavior of the real power system being modeled. To achieve this objective, an RTS needs to solve the model equations for one-time step within the corresponding wall-clock time. Therefore, it produces outputs at discrete time intervals when the system states are computed at certain discrete times using a fixed time step (Faruque *et al.*, 2015). Such simulators normally use discrete fixed time step simulation algorithms (Harley *et al.*, 1994) to solve the equations representing the simulated system (the model) with a constant calculation time at each time step. Some real-time circuit solvers include the capability to perform a limited number of iterations to increase the accuracy to simulate nonlinear equipment such as arrestors (Tremblay, 2012; Dufour *et al.*, 2017; Dennetière *et al.*, 2016).

To achieve this goal, RTS technologies use powerful computing platforms combined with optimized software, performant mathematical solvers, and special modeling techniques. An illustration of RTS architecture and the interactions between the RTS and the hardware under test is found in Figure 2.3.

The capability of an RTS to achieve real-time performance depends on various factors related to the hardware architecture and the specifics of the simulation platform. Some of these factors are as follows:

- simulation software and solvers optimized for real-time execution;
- high-performance parallel computation hardware (central processor unit (CPU), FPGA, memory, and fast computer cluster communication links);
- real-time operating system;
- fast real-time input–output systems to interface with devices under test (DUTs);
- interface with standard real-time communication protocols such as IEC 61850, C37.118, DNP3, and optical fibers;
- graphical user interface to control the simulation and view the results in real time;
- test management and result analysis; and
- model data management.

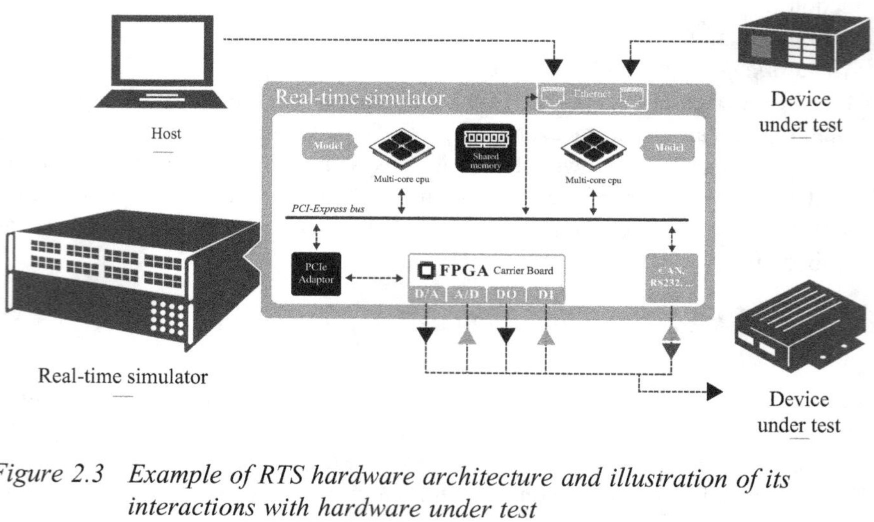

Figure 2.3 Example of RTS hardware architecture and illustration of its interactions with hardware under test

These aspects are discussed in more detail in Section 2.1.3, and they include the following:

- the type of simulation (TS or EMTs) and mathematical algorithms that solve the simulated system equations;
- mathematical models/modeling techniques suitable for real time; and
- the size and detail of the model under study, both of which significantly impact computation performance and the computational resources required to achieve the specified time step.

In short, RTS implementation requires highly optimized software, solvers, operating systems, and the capability to use several processors to execute the calculation in parallel in order to achieve the specified time step without overrun, regardless of the size and complexity of the model. The capability to scale up the processing power with the evolving system complexity is therefore critical.

2.2.4 Time constraints of RTS technologies

The timeline of operations occurring within a single time step is illustrated in Figure 2.4. To achieve real-time performance, all RTS operations must be completed within a **fixed time step T_s (where $T_s = t_n - t_{n-1}$)**.

The operations denoted as an aggregated function ***f(t)*** in the previous figure include the following:

- reading digitalized input signals from the input channels to the processor(s), sending signals to their corresponding port(s) in the model;
- executing the model solver(s), obtaining results corresponding to the dynamic response of the system under test or under simulation;
- exchanging signals between parallel processors (if any) to execute the model in parallel;
- sending output signals from the processor(s) to the output channels; and
- ... and some idle time, if required, because the start/end of each set of operations ($f(t_n)$, $f(t_{n+1})$, ...) at each time step (t_n, t_{n+1}, ...) must be synchronized with wall-clock time.

2.2.5 Accelerated simulation: faster-than-real-time and slower-than-real-time

As discussed earlier, an RTS must have sufficient computing resources to have idle time to allow calculations in real time. Since processors have limited operating clocks, at the time of this writing ranging from 2 to 5 GHz, multicore processors and computer clusters must be used to obtain real-time speed to prevent overruns, regardless of the complexity and size of the power system model. The RTS must then have the capability of splitting up the computational tasks to several processors to accelerate the calculation. Implementing efficient parallel processing is very challenging; however, RTS vendors and researchers have optimized parallel processing technologies over the last 25 years to enable the use of RTS with hundreds of processors (Weiwei *et al.*, 2018; Le-Huy *et al.*, 2017; Gagnon *et al.*, 2010).

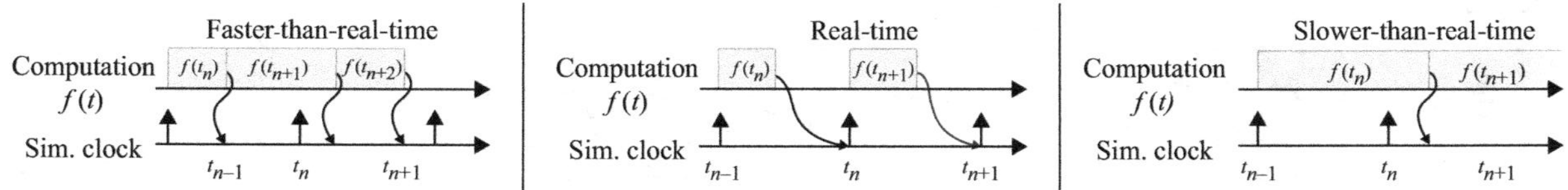

May be achieved with ***desktop/off–line simulation*** with small models.

The simulation is further ***accelerated with large and complex models using RTS*** fast parallel computing capabilities.

(a)

Strictly requires an RTS. Studies with or without interfacing with external equipment (HIL) can be performed.

The number of parallel processors, model decoupling or model optimization must be done until ***no overrun is detected***.

(b)

Usually the case with desktop/offline simulation. Slower-than-real-time

simulation of very large power systems using several processors can also be viewed as ***accelerated simulation as compared to desktop simulation***

(c)

Figure 2.4 Notions of (a) faster-than-real-time, (b) real-time, and (c) slower-than-real-time simulation

In addition to implementing parallel computation methods, the RTS must also have the capability to synchronize the computational processes themselves with an accurate hardware clock to ensure that the RTS is synchronized with the wall clock. In fact, the RTS must terminate all computation on the model within a period smaller than the specified time step to maintain sufficient margin to prevent over-run. Consequently, the RTS computer and software can also be used to accelerate the simulation to achieve faster-than-real-time simulation, if the necessity of synchronization with the hardware clock is removed. This operating mode is the same as any other simulation tools, such as EMTP or PSCAD, i.e., the simulation is performed as fast as the hardware permits.

RTS can thus be used to accelerate the simulation to achieve simulation speeds faster than real time, even for large power grids, if sufficient processors are used. If the number of processors is reduced, then the simulation speed will be slower than real time.

The simulation speed factor (*SF*) may be defined as

$$SF = \frac{\text{Simulation time}}{\text{Processing time}} \tag{2.1}$$

So, if the simulation time is 10 s and the processing time is 5 s, the *SF* is 2. If the processing time is 20 s, then the *SF* is 0.5. If the processing time is 10 s, then the *SF* is 1 or real time.

For small grids, most simulation tools will have *SF* greater than 1, i.e., the calculation time will be lower than the simulation time. However, an RTS can achieve an *SF* more than 1 even for very large power grids, including several power electronic systems such as HVDC grids. Even if the number of processors is reduced, for example, by 5, the RTS will achieve an *SF* of 0.2, i.e., the simulation speed will be five times slower than real time. However, conventional single-processor simulation software could reach an *SF* less than 0.01. This means that a 10-s simulation run could take more than 1,000 s. Of course, the SF does not always scale linearly with the number of processors but using optimized parallel computational methods accelerates the simulation as compared to single-processor simulation tools.

The importance of accelerated simulation in achieving simulation speeds that are faster than real time ($SF>1$) may become essential while producing the vast amounts of data required to implement and train AI-based controls. This acceleration also helps the deployment of multiple predictive simulations with DTs to help one to minimize the occurrence of blackouts, maximize resilience, and operate with seamless transitions and fail-safe methods at system restoration following severe events.

2.3 Introduction to the concepts of hardware-in-the-loop testing

HIL is widely used by utilities, manufacturers, and advanced research laboratories around the world for smart-grid-related applications (Montoya *et al.*, 2020). As mentioned previously, the main purpose of an RTS is primarily to conduct studies and testing of control and protection system equipment. The principle of HIL use during simulation is to "trick" the device or the piece of equipment interacting with the RTS into thinking that it is interacting with a real power system. The stimuli that are then produced by the simulated system provoke an instantaneous response from the control equipment in the loop, which, in turn, interacts with the simulator. So, HIL involves a closed-loop interaction between a piece of equipment (e.g., a power electronics controller, power plant excitation system, protection device, and actuator) and a model that is simulated in real time (a distribution system, wind turbine, solar photovoltaic (PV), motor drive, etc.). HIL is additionally a catch-all term with further subcategories.

In this section, we will focus on rapid control prototyping (RCP), CHIL, and power HIL (PHIL), as well as software-in-the-loop (SIL) simulation using manufacturer-provided equipment code to emulate their proprietary control systems. These various concepts of HIL are depicted in Figure 2.5, in which they involve the example application of solar PV inverters. SIL is explained later.

In all concepts of HIL, the illustration shows that the RTS interacts with external equipment through analog-to-digital (A/D) and digital-to-analog (D/A) converters that comprise the input and output interfaces of the RTS. RTSs are also equipped with discrete I/Os and with fiber optic communication interfaces to accommodate the exchange of large quantities of fast signals, for instance, power electronics gating signals. RTS also have Ethernet communication protocol interface capabilities, namely, C37.118 for phasor measurement units (PMUs), IEC

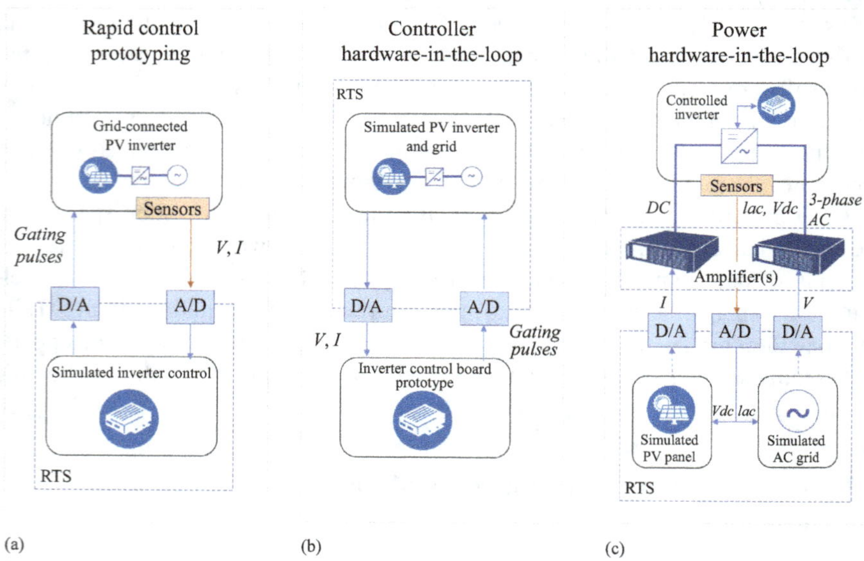

Figure 2.5 Concepts of HIL (a) RCP, (b) CHIL, and (c) PHIL

61850 protocols for substation protection devices, and DNP3 for SCADA communications as well as other protocols such as MODBUS.

2.3.1 Fully digital real-time simulation—a step before applying HIL techniques

Of course, before interfacing the RCP with the plant emulated in a second RTS using analog I/Os, it is a good practice to simulate the RCP and the plant in the same RTS. This is particularly useful if the power grid and the power electronic system are very complex. The RTS, including the plant and the RCP, will enable testing of the controller algorithm and protection functions for a very large number of contingencies in minimal time as compared to using a slower single-processor simulation tool. This step can be called "accelerated fully digital simulation." The interaction of the controller and power electronic systems under the analysis with other control and protection systems already installed on the grid can also be analyzed.

In practice, standard off-line fully digital EMT simulation tools can be used for this study phase for concept analysis, but fast parallel simulation is needed if hundreds of contingencies must be analyzed for very large power grids to perform more tests in less time or to perform hypothetical what-if scenarios.

2.3.2 RCP connected to a physical plant

In RCP, the RTS is used to model and simulate the control and protection algorithms in the early stages of design; so, it may be connected to a physical prototype of the controlled device (e.g., the PV inverter or motor drive) or to an actual plant. RCP has many advantages over direct implementation on production controller hardware. It is more flexible, faster to implement, and easier to debug than an actual microcontroller, which has limited interactive resources and must sometimes be programmed using a low-level programming tool. The modeled prototype can be tuned on the fly or completely and easily modified using the RTS graphical modeling engine. In many cases, the actual controller code can be generated directly for the final controller hardware using the RCP model to avoid implementing controller code manually and to decrease development time.

2.3.3 RCP connected to a real-time plant model through RTS I/O signals

The technique of interfacing an RCP to a physical plant as described earlier is very popular and efficient to develop power electronic controllers used in simple systems such as motor drives and low-power inverters. However, for complex systems, the RCP is connected to the plant model and also implemented with an RTS both through analog and digital I/Os. This method is very convenient for more complex plants that would be difficult or too expensive to implement at the early stage of the project. For example, implementing a physical bench that emulates a scaled-down MMC HVDC system connected to a grid model is quite complex and time-consuming. In such cases, it is preferable to simulate the complex plant in the RTS and to interface the RCP using I/Os. Once developed, the RCP may then be further tested on a scaled-down analog test bench, if necessary, before implementing the controller logic in the final controller hardware.

2.3.4 Controller hardware-in-the-loop (CHIL, or Often HIL)

In CHIL, the real controller (or a further advanced prototype of the controller) interacts with a simulated power system or renewable energy source. CHIL involves only interfacing low-voltage and low-power control and monitoring signals between the hardware controller (the hardware and the software under test) with the RTS. This is also known as "signal-level CHIL" or simply "HIL."

The same tests performed earlier during RCP project phases are normally repeated with the actual control hardware and software in HIL mode to verify that actual control and protection performance and behavior are the same as the prototype controller. This is very important since the controller code may be implemented manually and optimized to the same memory and processor resource as compared to the RCP system. The operating system of the actual controller may also be different. The results obtained with the RCP and digital simulation of the plant can be considered as the reference results, which must then be duplicated with actual controller hardware interfaced with the RTS in HIL mode.

2.3.5 Power-hardware-in-the-loop

In PHIL (Lauss *et al.*, 2016; Wang *et al.*, 2019), the RTS simulates a power system or power equipment that is connected to physical power equipment through an amplifier interface, the case in which the hybrid simulation is designated as an emulator or PHIL test bench.

PHIL test benches enable the circulation of power with nominal voltage and current through the DUTs to verify the performance under very realistic conditions. For example, PHIL benches can be used to test the thermal capability of a prototype DC–AC inverter (the DUT) by connecting the inverter to an emulated motor. The motor is emulated by a four-quadrant power amplifier controlled by an RTS ensuring that the amplifier current is identical to the actual motor current for various operating conditions.

Amplifiers are a key part of PHIL as the frequency range (or bandwidth), power capability, and voltage level of the testbed greatly affect the overall accuracy of the PHL test bench. One important technical aspect to consider is the capability of the amplifier to source and sink power in all four quadrants (P and Q, both positive and negative), especially for power system emulation, where the model in the RTS is a power system exchanging power with a bidirectional power source (e.g., a battery energy storage system). Not all amplifier technologies have symmetrical sourcing, and sinking characteristics and different technologies may have a more or a less significant output THD and bandwidth, which can affect the testing accuracy, depending on the application.

PHIL involves a closed-loop interaction and there may be stability issues due to the interface and overall loop delay caused by the amplifiers, the sensors, and the time to execute the model simulation. The interface/interactions are at the power exchange level, where phenomena under study can be very fast and may involve unnatural delays and latencies in the interface with the RTS (sensor time constants, amplifier response time, communication latency, etc.). Implementing accurate and stable PHIL test benches is a challenge.

2.3.6 Software-in-the-loop

SIL consists of simulating the complete systems using actual manufacturer equipment controller code provided by the manufacturer of the power electronic systems (PV farms, wind farm, HVDC, and FACTS). To the extent possible, the control system code provided by the manufacturer is an exact copy of the code implemented in actual controller equipment. The proprietary control code is, however, provided under confidential agreement and in the form of pre-compiled object code (DLLs), which is then interfaced with off-line or real-time simulation tools such as EMTP-RV, PSCAD, or HYPERSIM. SIL enables testing of the interaction between all controllers and the power grids, and analysis of the integration of new distributed generation and energy storage plants using control system models is very close to the actual system.

SIL is becoming very popular as it can decrease the time and the cost to test actual control hardware. Furthermore, it enables the analysis of the global performance of complex transmission, distribution, and microgrids integrating several wind and solar parks as well as HVDC and FACTS systems in fully digital simulation mode. SIL may also be used to implement power grid DTs and AI-based controllers using very accurate control models.

2.3.7 HIL as a standard industry practice: the model-based method and the V-cycle

All HIL testing techniques are widely used in industry by equipment manufacturers, research facilities, and power system operators as a powerful tool to develop quality products and better understand their interactions with the system that they are interacting with. In cases where it is used especially for product development, it significantly reduces time to market and it is an integral part of the product development lifecycle, typically depicted as a V diagram, as illustrated in Figure 2.6.

The V-cycle (Gausemeier and Moehringer, 2002) is a widely adopted conceptual standard in systems engineering and technology development. It is used for project planning and management, as well as for the quality assurance cycle for the development of very complex technology or systems engineering. It results in a much better project workflow than a typical "waterfall" (Ajam, 2018) project process does and considers the relationships of all development steps to one another.

The first series of steps found to the left of the V, in the downward flow, involves all activities relating to the project (or product) definition and conceptualization. In work involving simulation studies and prototyping, it involves

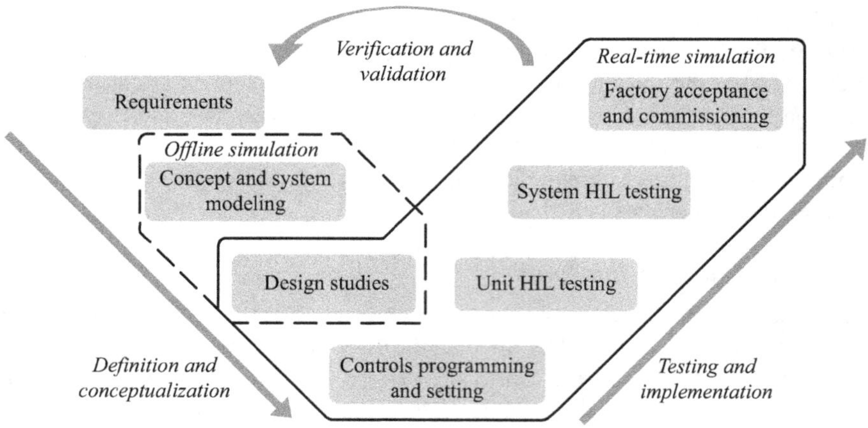

Figure 2.6 Technology development V-cycle involving simulation studies and HIL testing

system modeling and design studies, mostly using off-line or accelerated simulations, right up to the early concepts of control or algorithm programming. The second series of steps (the ascending flow to the right of the V) is concerned with testing and qualifying for commercialization and installation in the field by validating that the product or system responds as per design specifications and meets the design requirements. This is where all concepts of HIL are broadly used. Between the two branches of the V-cycle, there is the verification and validation (V&V) loop, which involves a series of steps followed iteratively, when unit tests, system tests, or factory tests require a reworking of the initial design or a customization of the product for a specific project.

The V-cycle concept ideally applies to the implementation of large HVDC transmission systems, interconnections, and distributed generation and power system modernization by utilities, as this effort requires many design steps, system studies, and most importantly acceptance testing.

For the most part, fully digital simulations are used for understanding the behavior of systems under certain circumstances resulting from dynamic interactions between power generators, loads, and their control systems. They help in defining the needs of the future grid according to load fluctuations and demand and further help in planning for the addition of compensation equipment or smart devices to better control the grid to meet security criteria.

On the other hand, HIL testing validates whether the selected equipment respects the specification and minimizes the financial and technical risks, as the utility engineer has sufficient proof to request any necessary design review and improvement by the manufacturer of the equipment under test.

A general mapping of all HIL concepts and the use of simulation across design stages is depicted in Figure 2.6. Of course, there is no exclusive use or restriction of any of these engineering, research, and design techniques in differing application spaces, but this is an accurate view of the typical current usage (Table 2.1).

Table 2.1 General application spaces of fully digital simulations and HIL testing in smart grid

	Research	**Design and planning**	**Prototype**	**Test**	**Operation and maintenance**
Rapid control prototyping (RCP)	✓		✓		
Controller hardware-in-the-loop (CHIL)	✓		✓	✓	✓
Power hardware-in-the-loop (PHIL)	✓		✓	✓	
Fully digital simulation and faster-than-real-time simulation	✓	✓			✓

Note that all RTS-based power system simulation and testing techniques are used throughout research in general. As a matter of fact, they are one of the most flexible technologies allowing translation of concepts and algorithms from the mind of productive researchers to the physical laboratory space, including novel power electronics topologies, controls, and algorithms. Accelerated and faster-than-real-time simulation offers, for instance, the possibility of producing a large volume of intelligible data, based on realistic system models, that can then be used to train AI algorithms—the implementations of which are to be tested downstream using HIL.

Regarding the operations and maintenance application space, CHIL and fully digital simulations, including SIL, have been used with validated accurate models of the power system and its equipment. Simulations are used to support decision-making, when important maneuvers or reconfigurations of the system are considered but known to have potential adverse effects on the continuity of service and system stability. On the other hand, CHIL is used to test replicas of controls and protections installed in the field for post-contingency assessment, settings change acceptance, and even future improvements of equipment controllers.

2.3.8 Bandwidth, model complexity, and scalability considerations for RTS applications

Let us briefly review basic system dynamics before getting into the simulation aspects. Figure 2.7 depicts a mechanical and an electrical system. The mechanical system consists of a rotating mass defined by its moment of inertia J moved by an external torque T, a spring (or flexible shaft) with a constant k and a damper (or friction) with a damping constant b, whereas the electrical system consists of a voltage source V connected in series with a resistance R, an inductance L, and a capacitor C.

Both the systems are described as damped harmonic oscillators in the literature (Nise, 2019) and have dynamics expressed as a second-order differential equation. The output variables of interest of each system are the rotary position of the mass θ

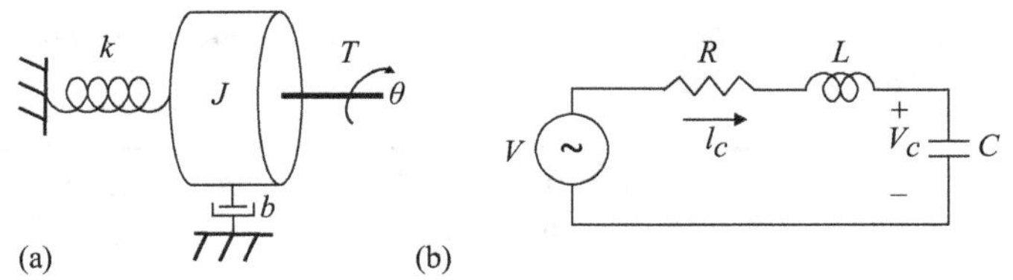

Figure 2.7 Examples and mathematical expressions of second-order: (a) mechanical and (b) electrical systems

and the voltage of capacitor V_c, respectively. The dynamic properties of both the systems can be expressed in terms of their natural frequency ω_n in rad/s (or resonance frequency f_n in Hz) and damping factor ζ. By observing the analytic form of their damping factors, one can observe that it is directly proportional to b for the mechanical system and directly proportional to R for the electrical system. Moreover, the natural frequency (rate of oscillation) of each system is respectively a function of the spring-mass ratio and the inductance-capacitance product.

$$T = J\ddot{\theta} + b\dot{\theta} + k\theta \tag{2.2}$$

$$v = LC\ddot{v}_C + RC\dot{v}_C + v_C \tag{2.3}$$

$$\omega_n = \sqrt{\frac{k}{J}} \tag{2.4}$$

$$\omega_n = \frac{1}{\sqrt{LC}} \tag{2.5}$$

$$\zeta = \frac{b}{2\sqrt{kJ}} \tag{2.6}$$

$$\zeta = \frac{R}{2}\sqrt{\frac{C}{L}} \tag{2.7}$$

$$f_n = \frac{\omega_n}{2\pi} \tag{2.8}$$

For simulation, the modeling principles are analogous, but one key parameter dictating the selection of simulation time step value T_{step}, or the model sampling frequency F_s ($1/T_{step}$), will be the frequency range for which the simulation result accuracy is expected.

In the previous examples, assuming the mechanical example represents a large power plant turbine-alternator; the inertia is such that oscillation modes of the generator will have periods in the order of seconds; thus resonance frequencies are well below the nominal frequencies of the electric system (typically 60 or 50 Hz), from less than 1 Hz up to a few Hz, according to (2.4) and (2.8).

Theoretically, the model sampling frequency should be at least twice the value of the simulation's highest frequency of interest. But in practice, as of rule of thumb, the model sampling frequency should be at least five to ten times more than that. For example, simulating mechanical oscillation modes of the generators as discussed earlier, typically ranging from 0.5 to 2 Hz, depending on the total turbine-alternator resulting inertia, requires a simulation time step ranging from 50 to 200 ms. In practice, experts analyzing rotor angle oscillations in large power systems will use time steps ranging from 1 to 5 ms depending on mathematical solver accuracy and other control systems simulated. These latter factors may include voltage and speed regulators as well as power system stabilizers with smaller time constants that will affect the dynamic performance of the electro-mechanical systems. Rotor oscillation frequencies of large multi-machine systems will depend on energy transferred between machine and loads, which depends on fundamental frequency, impedances of machine, transformers, and lines. The detailed electromagnetic models of transmission lines, transformers, and other components of the grid are therefore not required.

On the other hand, if study objectives require evaluation of amplitude of overvoltage and overcurrent induced by faults and breaker operations, then detailed EMT simulation models are required to simulate fast transients. As shown in (2.5) and (2.8), resonance frequency f_n of a simple R–L–C circuit depends on the values of L and C, while R mostly influences damping. For simple cases, the voltage source and the R–L circuit of Figure 2.8 can be seen as the Thevenin equivalent impedance and is often calculated using the short-circuit power S_{sc} (frequently given in MVA) at the point of common coupling (PCC). Assuming that R is much smaller than the inductive impedance $X_L=2\cdot\pi\cdot f_1\cdot L$, the equivalent inductance can be estimated, for a three-phase system, as follows:

$$S_{SC} = \frac{V_{rms\ L-L}^2}{2\pi f_1 L} \tag{2.9}$$

where f_1 is the fundamental frequency and $V_{rms\ L-L}$ is the Thevenin equivalent voltage. The resistance R is often given from a ratio of X_L over R. The resonant frequency f_n induced by L and C can easily be estimated as

$$f_n = f_1\sqrt{\frac{S_{SC}}{S_C}} = f_1\sqrt{\frac{X_C}{X_L}} \tag{2.10}$$

where S_C is the reactive power of the capacitor at nominal voltage, X_C and X_L are the reactance of the capacitor and that of the equivalent inductor, respectively, at fundamental frequency. For example, on a 50-Hz power system, if a 100 MVar three-phase capacitor bank is switched at a bus of the network in which the three-phase short-circuit power is 6,000 MVA, the frequency f_n of the inrush current transient (or capacitor bank energization current) will be approximately 387 Hz. If the size of the capacitor bank is 35 MVar and the short-circuit power is 15,000 MVA, the resonant frequency is approximately 1.04 kHz. Simulating the first example would require a sampling frequency of about 2–4 kHz (or a T_s of 250–500 μs), while the latter example would require a sampling frequency of about 5–10 kHz (or a T_s of 100–200 μs).

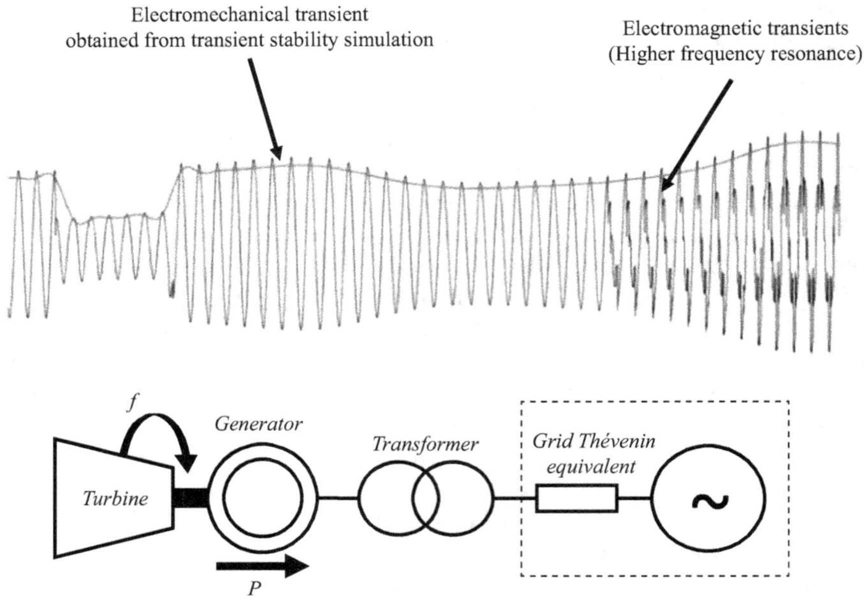

Figure 2.8 Differentiation of transient stability and electromagnetic transient simulation domains with an example of power generator and grid phenomena

Electric system resonance frequencies may also be much lower in other cases. The first example is when long transmission lines equipped with large shunt inductors to reduce the charging current at "no-loads" are switched off. For example, a 60-Hz, 300-km, and 765-kV transmission line has a no-load charging reactive power (capacitive current of the line at no-load) of about 700-MVar and is compensated using two shunt reactors of 330 MVar each. The resulting resonance frequency is 58 Hz, that is, very close to the fundamental frequency. If only one shunt reactor is switched off, the resonance frequency is 41 Hz. In some cases, very long transmission lines must be compensated with series capacitors to increase their power transfer capability, often leading to resonance frequencies of a much lower value than the system operating frequency. Considering the compensation factor given by the ratio of X_C over X_L, a compensation factor of 50% on a 60-Hz system will lead to a resonance frequency of 42 Hz, whereas a compensation factor of 25% will lead to a resonance frequency of 30 Hz. Such low electric system resonance frequencies can excite torsional vibration modes of thermal generators with long shafts and damage them.

In practice, impedance of transmission lines as a function of frequency exhibits a series of poles and zeros, which will be excited during faults and breaker operations (energization and de-energization). Transmission line switching will produce voltage and current transients with the shapes close to a square wave due to travelling wave effect, and fast transients with risetimes in the order of 50–200 μs will occur depending on the line length and system short-circuit power. Therefore, in practice, transient frequencies expected in electromechanical systems can range from below 1 Hz for mechanical oscillations to a few kHz for electromagnetic phenomena. Simulation time step values required the range from a little less than 10 μs and up to 4–10 ms.

Since modern power systems are increasingly complex and include a large amount of power electronic components integrated within the conventional power system, it is important to consider the best tools and configuration for RTS studies and HIL testing. One must first identify the purpose of study and then evaluate the expected transient frequencies and subsequently the type of simulation and modeling techniques required for the application. For every type of RTS platform, optimized software and hardware are essential to successfully meet the requirements for real-time simulation.

2.3.9 Transient stability and electromagnetic transient simulation methods

Simulation is a powerful tool for understanding the deepest inherent particularities of a system from the basics to complex advanced technologies and larger systems. Figure 2.8 shows an example that allows the distinction or classification of some power system phenomena and the introduction to the use of different simulation methods, namely, the TS (electromechanical phasor simulation method) (Kundur *et al.*, 2004) and the EMTs method (Dommel, 1997).

Without going into the details of the waveform (magnitudes, variable monitored), one can observe various perturbations. Under normal conditions, the generator rotates at a nominal system frequency (60 or 50 Hz) and produces a sine wave back-EMF of constant amplitudes. It provides nominal active power corresponding to the governor (turbine controller) power set point and absorbs or produces reactive power, according to the excitation system (generator terminal voltage control) control mode and set point. During steady-state conditions in a power grid, there is a balance between energy generation and consumption. Any event that disrupts this balance, such as load shedding or transmission line outages, has an impact on the synchronous generator speed and consequently the system's frequency. In such scenarios, while the local or wide area controllers will try to regulate the frequency back to its nominal value, dynamics in the system will cause stable or unstable power swings. Although these electromechanical oscillations (significantly slower than EMT phenomena) can be simulated using EMT-type tools, they can also be accurately represented using electromechanical TS tools (TS-type tools) based on the phasor method, as long as there is no interest in the analysis of signals with frequency contents higher than the fundamental frequency of the grid.

On the other hand, a sudden energization of the transformer, switching of capacitor banks, or transmission lines will cause fast transients with higher frequency contents, provoking oscillations in the order of kHz, namely, EMTs. These phenomena can be simulated using EMT simulation, but not with TS tools. It is important to recall here that TS tools represent the power grid with a simplified model that is nevertheless valid at the electric system fundamental frequency and below, using differential algebraic equations to calculate the dynamic flow of power between machines and loads. The dynamics of the system (poles and zeros) are mostly dictated by machine inertial response and rather slow control systems. Furthermore, most TS tools used for transmission systems represent balanced systems using only the positive sequence representation of the electric circuit. With an RTS using TS tools, a single processor core can simulate systems with several thousands of buses, faster than real time, using a time step of approximately 20 ms. On the other hand, EMT simulations require simulation of the detailed dynamic of the transmission systems, including all poles and zeros induced by inductors and capacitors spread over the generation, transmission, and distribution systems. The simulation time step value will additionally range between 10 and 100 μs to simulate high-frequency oscillation up to a few kHz. Modern RTS will therefore need parallel computer systems with several high-end processors to reach real-time capability.

A list of simulation software and platforms is given in Table 2.2.

Typically, an RTS using an EMT-type solver requires a time step of roughly 50 μs for the simulation of most transmission systems, and time steps in the range of 10–50 μs for the simulation of distribution systems, depending on the technologies involved in the modeling as well as the system size, in order to faithfully simulate transients.

For applications involving fast power electronic applications (drives, inverters), the required simulation time step could go as low as a few hundred nanoseconds, for instance, in certain drive applications, in which case the RTS processors must be much faster than commercially available CPUs. FPGAs are therefore commonly used for such simulations.

Figure 2.9 illustrates the simulation software and hardware involved in RTS applications.

Conversely, in the case of TS-type simulation, a time step in the range of 1–20 ms is sufficient for the RTS solver to capture the TS dynamics caused by electromechanical phenomena in the system. However, the solver must be both flexible and optimized to take the advantage of parallel-processing-based techniques, since the size of the nodal matrix to get factorized augments with the number of nodes existing in the power grid. An example of simulating a confederated transmission and distribution system that involves more than 108,000 nodes with a time step of 10 ms and exploiting nine parallel processors is presented in the study of Jalili-Marandi and Bélanger (2018). Such performance is obtained by splitting the global systems into several smaller subsystems interfaced with a Thevenin equivalent.

The RTS computing burden in the context of power systems can be characterized in terms of the number of system nodes (or buses) in the simulation as well as the time step required for accurate simulation. One could simplify the factors involved in the real-time calculation burden of a simulated system (for the purpose of comparing two system simulations) as a qualitative index *RTb*, which

Table 2.2 Short list of commercially available and open-source simulation software/platforms

	Off-line	RTS
TS	PSS/e (Siemens PSS/e PTI, n.d.), PSLF (PSLF | Transmission Planning Software | GE Energy Consulting, n.d.), TSAT (TSAT^{TM}—Powertech Labs, n.d.), PowerFactory (PowerFactory—DIgSILENT, n.d.), CYME (CYME Power Engineering Software, n.d.), ETAP (ETAP | Electrical Power System Analysis Software | Power Management System, n.d.), PowerWorld (PowerWorld » The Visual Approach to Electric Power Systems, n.d.), NEPLAN (NEPLAN—Power System Analysis, n.d.), ePHASORSIM (Jalili-Marandi *et al.*, 2013), GridPACK (GridPACK, n.d.), PSAT (PSAT, n.d.)	ePHASORSIM (Jalili-Marandi *et al.*, 2013)
EMT	EMTP-RV, PSCAD/EMTDC, PowerFactory, ATP	OPAL-RT HYPERSIM, RT-LAB/ ARTEMIS, RTDS
EMT (FPGA)		OPAL-RT eFPGASIM (eHS), TYPHOON, Speedgoat, dSPACE

(All product names, logos, and brands are property of their respective owners. All company, product, and service names used in this chapter are for identification purposes only.)

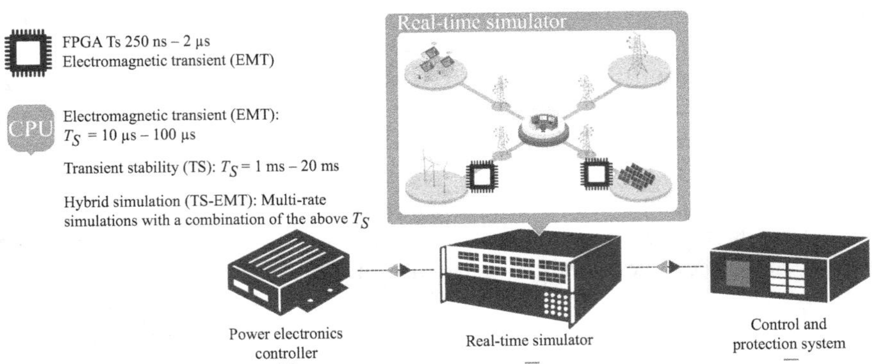

Figure 2.9 A real-time simulator connected to external hardware—an illustration of simulation software and hardware components involved in RTS applications

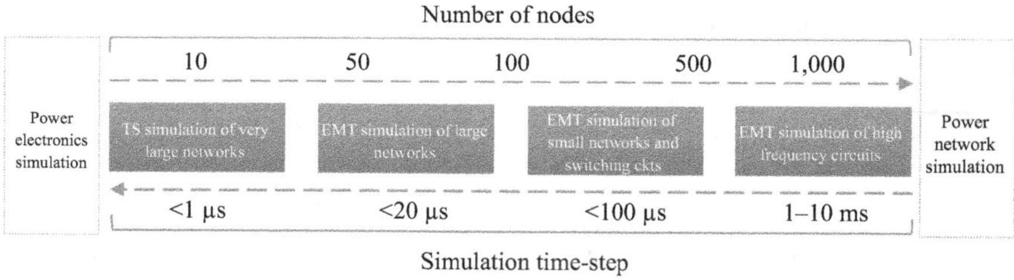

Figure 2.10 Illustration of (a) application space and (b) typical use cases of TS and EMT simulation, as a function of a number of system model buses (n_{busses}) and time step (T_s)

would be the ratio of the number of buses (n_{busses}) to the simulation time step (T_s). Accordingly, the *RTb* index for an RTS increase proportionally to n_{busses}, corresponding to an increase in the time it takes to achieve model calculation operations within a single time step. The *RTb* index will also increase as the required T_s decreases, resulting in the acceleration of the RTS clock and a shorter amount of time to achieve calculation operations. Figure 2.10 depicts the mapping and the relationship between T_s, n_{busses} and applications of interest.

One must, however, note that using the simple index *RTb*, itself based on the ratio of the number of buses (n_{busses}) to the simulation time step (T_s), is an approximation that disregards the complexity of the circuit connected to each bus. It should be obvious that a station containing several transformers, transmission lines, HVDC, FACTS, and loads will take more processing time to compute than a simpler station containing only two transmission lines. Furthermore, the number of breakers connected to the same buses will also affect the processing time, as the nodal matrices must be re-inversed at each modification of breaker status. Consequently, realistic benchmarks must be performed to evaluate the exact number of processors required to achieve specified time steps. RTS manufacturers can also provide more accurate methods to evaluate the number of processors as a function of the quantities and types of network components to simulate.

Typically, equipment-level power electronics simulation requires smaller T_s but involves a smaller n_{busses}, whereas large system simulations involve larger n_{busses} and (nonexclusively, however) larger T_s. When the size of the system simulated in the RTS increases to a very large n_{busses}, in excess of 500–1,000 buses, detailed EMT simulation is usually not required, as the overall system electromechanical stability is the main focus of the HIL tests or simulations, as opposed to the fast EMT phenomena and the detailed switching mechanism of power converters. Where the best of both simulation domains is required, a hybrid TS-EMT simulation can be performed, which is quite an advanced technique still under research and investigation by academics, research labs (Jalili-Marandi *et al.*, 2009), and RTS manufacturers (Jalili-Marandi and Bélanger, 2020).

One additionally important aspect to consider is the scalability of an RTS and the model it simulates. Discretized system equations have states and output variables dependent on states at the current time step and states at the previous time step. While dividing circuit equations into separate processes, it creates an algebraic loop (because some states become unknown to the difference equations). Algebraic loops must be broken using a discrete time delay (or a one-time step delay).

A classical approach used for parallelizing power system circuit equation calculations is to take the advantage of the intrinsic wave propagation delay τ of transmission lines, which is a function of its characteristic capacitance C (expressed in Farad per unit length), inductance L (expressed in Henry per unit length), and line distance (or length) d, as given in the following equation:

$$\tau = d\sqrt{LC} \tag{2.11}$$

In practice, for most overhead transmission lines, one can approximate the propagation delay τ as being proportional to the inverse of the speed of light, time, and the distance d: $\tau = 1/c$ (where $c \approx 3 \times 10^8$ m/s$\approx$300 km/ms), but this rule of thumb is not applicable to all cases, especially to cable systems where capacitive charging is considerably larger and, as a result, the propagation delay is also larger. As an example, an overhead transmission line with a length of 300 km will have a wave propagation delay of τ = 300 km/(300 km/ms) = 1 ms. Similarly, a 15-km line will have a propagation delay of 50 μs. This means that subsystems separated by lines longer than 15 km can be solved in parallel without inducing any error, since the time step is smaller or equal to the propagation delay of the line.

Consequently, mostly, if each real-time simulation software does not have this decoupling technique implemented using the Bergeron transmission line model, it is also known as the distributed parameter line, adapted for model separation (Dommel, 1969) and essential to implementing efficient parallel processing. As previously outlined, the prime constraint of this model, however, is that it can only be simulated if the time step T_s is smaller than or equal to the propagation delay τ of the line. This approach has no adverse effect on the simulation accuracy and does not affect numerical stability.

Another well-known approach is the use of stubline model (Rivard *et al.*, 2018) to decouple the electric system equations involving shorter lines or cables through a decoupling inductance, or a decoupling capacitor. The schematic diagram of a single-phase stubline is depicted in Figure 2.11.

The stubline is equivalent to a line modeled with a resistive–inductive–capacitive (RLC) PI circuit but with the wave propagation and parametric constraints that d = 1 and $\tau = T_s$. In other words, while modeling, for example, a decoupling inductance, the stubline adds a parasitic capacitance that is a function of the time step, leading to the introduction of a resonant pole at $1/T_s$ Hz. This adds certain inaccuracies (in the form of numerical oscillations; Marti and Lin, 1989) during transients, if this pole is dominant and undamped as compared to the rest of the simulated system's poles. The parasitic capacitance also modifies reactive power in steady state. While using stublines, generally the larger the inductance used as a decoupling element, the less dominant the parasitic effect of the added capacitance

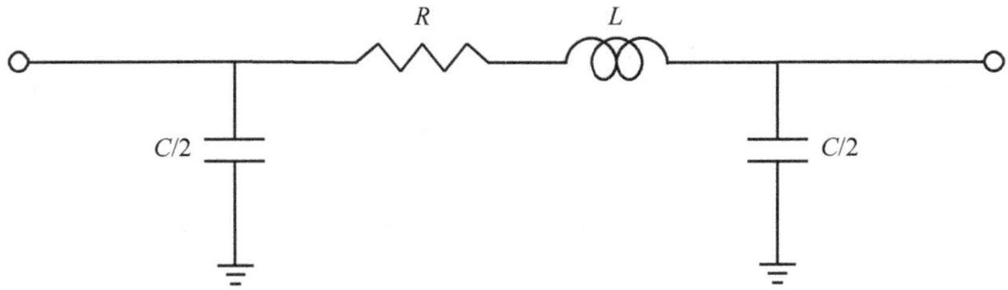

Figure 2.11 Schematic diagram of a single-phase PI section line model (stubline is based on the principle of a PI section)

will be. As an extension of this idea, the smaller the time step of the simulation, the lesser the inaccuracy caused by the parasitic component of a stubline decoupling will be. The same idea can be used to simulate a decoupling capacitor. In this circuit, the sum of the parallel capacitors (equals C) is equivalent to the simulated capacitor. However, a parasitic inductor (with size that depends on C and T_s) is added between the two capacitors, which adds a resonant frequency not included in the actual circuit.

Other decoupling techniques are used and described in the literature and they are applied to decoupling digital models as well as interfacing techniques for PHIL (Lauss *et al.*, 2016). One common decoupling technique of note is the ideal transformer method (ITM). It is the most used common decoupling technique outside of the transmission line type decoupling because of its conceptual simplicity, as illustrated in Figure 2.12. Such decoupling techniques can be very useful, but they need to be applied carefully because they can cause numerical instability. Although not exclusively required, this technique is preferred for decoupling DC circuits or circuit portions with lower signal bandwidth. Of course, the lower the T_s and the lower the added discrete time delay, the lesser the risk of numerical instability will be. The ITM method also introduces parasitic effects from the added numerical delays, like the stubline method, but the stubline method is intrinsically stable, while the ITM can be numerically unstable if T_s is too large for the application.

Another important decoupling technique uses Thevenin or Norton equivalent networks to interface several subnetworks solved independently in parallel. The state-space nodal (SSN) method (Dufour *et al.*, 2010, 2011), commercially available through OPAL-RT's RT-LAB/ARTEMiS software, can, however, significantly increase the size of the power systems that can be simulated at a lower time step, without adding artificial delays or parasitic components. One major advantage of this solver is the capability to decouple complex system models with short lines, without the need to reduce the time step to match the line propagation delay to use decoupling transmission lines or without the need to use stublines. In real-time applications, another noticeable advantage is that the method decouples the switches into different groups, making their precalculation simpler and thus lowering the memory requirements when compared to the full precalculation of switch

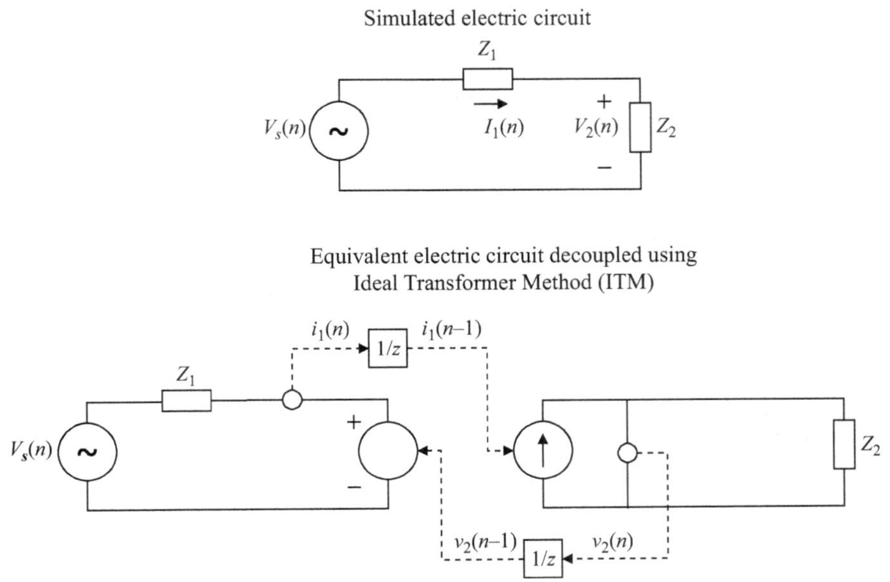

Figure 2.12 Illustration of the ideal transformer method (ITM) decoupling used in parallel real-time simulation

permutations in the standard state-space method. As depicted in the single-line diagram of the network model of Figure 2.13, the model equations can be decoupled at given nodes of the system and calculated in parallel.

The algorithms involved in such solvers are not trivial, but the principle of SSN can be rendered as follows. Assuming that the whole power system model (including Groups 1 and 2 depicted in Figure 2.13) is represented using state space and computed using a state-space solver such as Simscape Electrical (Simscape, n. d.) or ARTEMiS (Dufour *et al.*, 2013), it is then generally represented as in the following equation:

$$x_{n+1} = \begin{bmatrix} A_k \end{bmatrix} x_n + \begin{bmatrix} B_k \end{bmatrix} u_{n+1} \tag{2.12}$$

Equations can be rewritten as two separate state-spaces as in (2.13), their respective calculations can be parallelized and common node information is obtained from the nodal equation of (2.14) in the form of injected variable within the u_{n+1} vector of each state-space.

$$x_{n+1} = \begin{bmatrix} A_{1_p} & \\ & A_{2_q} \end{bmatrix} x_n + \begin{bmatrix} B_{1_p} & \\ & B_{2_q} \end{bmatrix} u_{n+1} \tag{2.13}$$

$$i_{g1_{n+1}} + i_{g2_{n+1}} = Y \cdot v_{node_{n+1}} \tag{2.14}$$

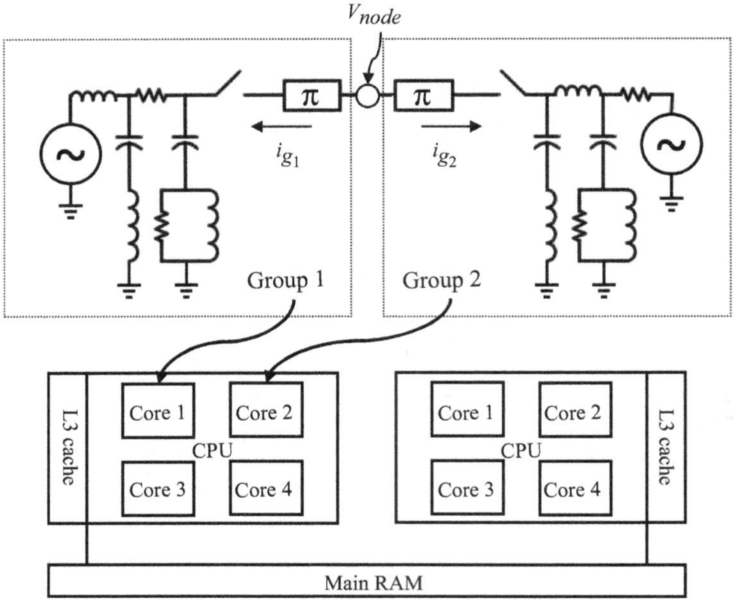

Figure 2.13 Illustration of the principle for the ARTEMIS-SSN decoupling method

Several solvers have been developed over the last 20 years using this principle, such as MATE (Martí *et al.*, 2002) and GENE (Strunz and Carlson, 2007). Such methods do not add errors to the simulation but require more data communications within the same time step, which limits the efficiencies of parallel processing.

The implementation of efficient circuit solvers taking full advantage of parallel processing is still an art. Besides the mathematical formulation discussed earlier used to decouple large systems into several independent subsystems, the implementation of the software must be efficient and optimized for specific multicore processor architectures and distributed computer clusters. The inter-processor communication overhead must be maintained below 1%–5% of the targeted time step—where it becomes a challenge to achieve time step values below 50 μs. Poor software implementation may lead to inefficient calculations, which, in turn, can lead to computationally inefficient parallelization, as more processors are added and inter-processor communication increases.

Other challenges are the capability to automatically and optimally allocate the calculation of subsystems to each available processor unit as well as to enable fast communication to external equipment for HIL simulation. However, RTS manufacturers have managed with these challenges and taken advantages of modern multicore processor technologies. Such progress, having been achieved over the last 20 years, now enables the building of DTs of large power grids integrating significant quantities of wind and solar parks, requiring the implementation of AI-based distributed control systems.

2.3.10 Smart-grid testbed attributes and real-time simulation fidelity

RTS-based power system testbeds can be compared with hardware testbeds and actual power system installations in terms of three main attributes: their cost, the fidelity of the tests they perform, and the coverage (or range, or variety) of tests that can be performed (Salcedo *et al.*, 2019). As illustrated in Figure 2.14, these attributes are specifically compared to fully digital simulations (SIL, using manufacturer equipment controller code), CHIL (using actual control equipment), and PHIL laboratories using full-scale or downscaled equipment and actual systems (or the real power system during commissioning tests). Note that RCP has not been included for the sake of simplicity, but that it can be considered here as a subcategory of CHIL.

Of course, cost may depend on the complexity and purpose of the testbed, but it is undeniable that the cost and the risk of the physical power system or power plant are much higher than that of fully digital simulation, CHIL, and PHIL. Furthermore, the test coverage of using an actual power grid is much less, since the critical and worst case-operating conditions that must be prepared for and faults cannot easily be tested or are impractical to test.

Fidelity may be interpreted in various ways, but in this figure, it is designated as "how closely the testbed approaches realistic conditions" to the DUT or for the purposes of the test. It is worth noting that fully digital simulation normally offers the same simulation fidelity as the RTS used in a CHIL or PHIL setup, but here it is qualified with a minus (−) because it cannot interact with a DUT. However, fully digital simulation can increase the accuracy and test coverage as compared to CHIL, since interaction with several control systems can be investigated, which can otherwise be more expensive if all controllers in the power system are to be represented by a replica of actual control equipment. This technique has been adopted by a few large utilities in China. The fidelity of a fully digital simulation may, however, be improved by representing each controller with their manufacturer's controller code, for instance, for wind and solar plants as well as HVDC and FACTS. As outlined previously, this is also referred to as SIL. Power grid DTs used in AI-based techniques to control global grid performance and reliability will certainly take the advantage of SIL.

SIL will also enable the monitoring of local equipment control performance by comparing the measured control response with the response expected from the manufacturer's equipment controller code. It will then be possible to detect improper adjustment of control systems as opposed to manufacturer or system operator recommendations.

The fidelity of downscaled equivalents is debatable, because it depends entirely on the parts composing this testbed and additionally depends on the final purpose. In general, downscaled equivalents are great to use for prototyping or proving design concepts, although they do not necessarily allow mimicry of the real application of the power system with high fidelity. On the other hand, RTS-based

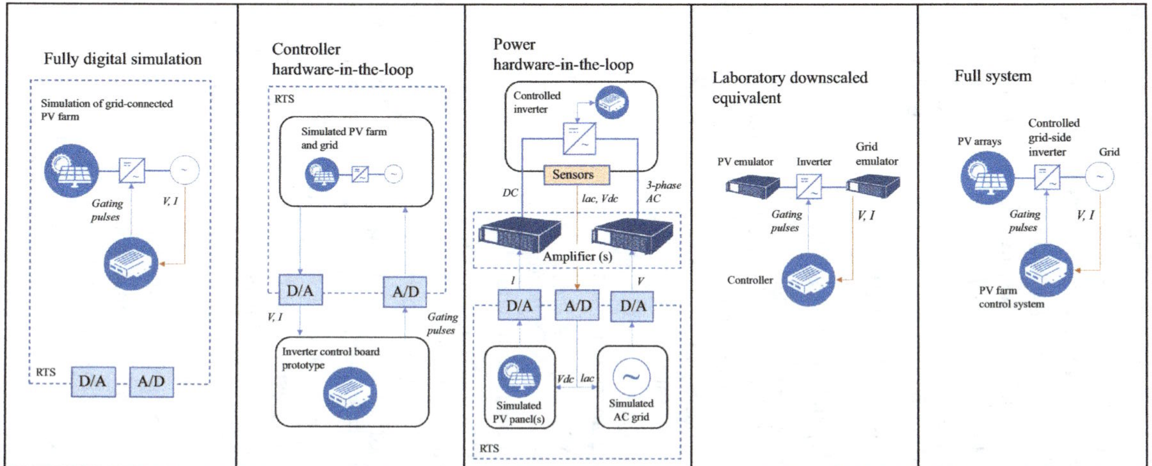

	Fully digital simulation	Controller hardware-in-the-loop	Power hardware-in-the-loop	Laboratory downscaled equivalent	Full system
Cost	–	+/–	+	+	+++
Fidelity	–	++	++	+/–	+++
Coverage	+	+++	++	+/–	–

Figure 2.14 Qualification of power system testbeds and real system testing in terms of cost, test fidelity, and test coverage

testbeds using validated and high-fidelity models can very accurately mimic real system conditions and cause the DUT to interact accordingly.

Finally, the test coverage is an aggregate measure of how many tests were performed, which kinds of tests they were, and how deep and repeatable the tests are with the various testbeds. With validated and accurate models, fully digital simulation is a powerful tool for design studies, and multiple tests can be run in batch, potentially faster than real time. The fact that it is a simulation unlocks the possibility of simulating scenarios that would otherwise be impossible to do on the physical power system because of the introduction of faults or their destructive nature. Therefore, in terms of device testing, CHIL provides the best of overall test coverage for testing protection systems and controls, allowing fault scenarios, transitions, and dispatching scenario functional testing in a closed loop with realistic behavior. SIL would extend the capability of CHIL by representing simulated controllers with the respective manufacturer's equipment controller code, while other critical controllers would be implemented using actual hardware replicas. The integration of SIL and CHIL for critical controllers may be considered as the ideal solution in terms of cost, scalability, accuracy, and test coverage, when the objective is to test the performance of the global grid.

PHIL also allows fairly representative test coverage as some (yet not all) of the simulated faults can be seen by the DUT through the amplifier interface, but depending on the voltage, maximum current, bandwidth, configuration, and power characteristics of power amplifiers, the test coverage is limited as compared to CHIL. Tests performed on the real power system are essential and a part of all engineering projects at the commissioning phase. These are the final tests applied to approve the installation as conforming to design specifications.

The high fidelity of RTS-based testbeds will most likely depend on the application of essential guidelines. The RTS needs to be (numerically) stable and the simulated system must be accurate in the range of the phenomena of interest. In general, the accuracy and stability are achieved with the selection of the sufficiently small time step T_s. Many advanced concepts explain the stability margin of a simulation, but as a general guideline, the smaller the T_s is, the larger the stability margin. Accuracy is also better as T_s becomes smaller, down to a value satisfying the simulation bandwidth, and is a function of various concepts that include modeling techniques and solvers. In general, T_s must be large enough to achieve real-time performance and small enough to ensure acceptable accuracy and stability. As illustrated in Figure 2.15, achieving acceptable fidelity from real-time simulation is a compromise, or trade-off, between computation burden and achieving a numerically stable and accurate simulation. "Trade-off" in this usage implies that as one improves, the other suffers, and so there is an optimal range located somewhere between the two extremes.

In terms of power system model accuracy, the guideline is usually given in terms of the highest resonance frequency f_r (or highest frequency) of interest in the simulation, where $T_s{\cdot}f_r$ <5%–10% (ex. T_s = 25–50 μs for a transient of f_r = 2 kHz). For power electronic converter simulation, the guidelines are a function of the switching frequency f_{switch}, where $T_s{\cdot}f_{switch}$ <0.2%–1% (ex. T_s = 2–10 μs for

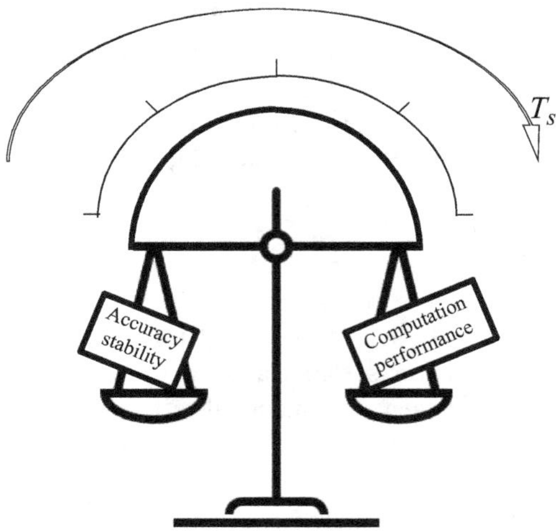

Figure 2.15 The RTS-based testbed simulation fidelity compromise

f_{switch} = 1 kHz). Special converter models called switching function models (Terwiesch *et al.*, 1999; Dufour *et al.*, 2020) will, however, offer better accuracy under certain conditions with $T_s{\cdot}f_{PWM}$ <10%–20%. In any case, acceptable testbed loop accuracy is achieved if the latency of input/output channels to interact with the DUT is smaller than 10% of the control system sampling period. This is particularly important for fast inverter control and protection system testing.

2.3.11 Importance of model data validation and verification

In the modeling phase of an RTS-based testbed, model data V&V is an important aspect of ensuring stability and accuracy. As a best practice, results obtained with real-time models must be compared with off-line simulation results using smaller (Dufour *et al.*, 2020; Terwiesch *et al.*, 1999) time steps or variable step solvers as a reference. The V&V simulation test should also include scenarios in various steady-state conditions and transient events able to cover different transient phenomena of interest.

One important aspect is the model data verification. Incorrect data may not be detected by comparing results obtained with different solvers if the same word data are used. A good practice is to verify each model parameter against typical system data for the same voltage level. Some other parameters can also be recomputed to compare with physical data such as the traveling wave speed of overhead transmission lines, which should be between ~98% and 99% of the speed of light for the line-mode (positive sequence) and a lower speed for cable systems. Load flow results using RTS can also be compared with results obtained with TS simulations and load flow software used in planning and operating departments. Using a centralized database accessible to every software, including RTS, is also a good practice to minimize data entry errors that are likely to occur in the case of large power grids.

2.3.12 Test scenario determination and automation

One of the main challenges is to design test plans for specific studies to ensure the complete test coverage to reach study objectives. Systems must be tested not only during steady-state operating conditions but also for small system disturbances and during balanced and unbalanced fault to verify system stability and stresses on various components. These tests must be repeated for several system contingencies and loading conditions. The number of test cases to perform can quickly become large and impractical in some cases. Consequently, the test plan must be optimized to reach study objectives within the allocated time and budget.

For small studies, tests can be executed manually. The specialist can execute a small series of tests, analyze the results, and prepare the next test series based on previous test results to find critical cases as quickly as possible. For larger studies with several contingencies, the specialist can program scripts to automate the test execution of several test series in parallel. Again, the next batch of tests to execute will be selected and launched after the analysis of previous results.

However, the number of tests to perform and analyze can become impractical for large and complex studies with hundreds of contingencies, various fault types, duration, and location of events. The determination of power transfer capability of a large power system under normal and abnormal conditions for $N-1$ contingencies (meaning one network element is out of service) may lead to the execution of several thousands of simulations. In such a case, the AI-based automated process illustrated in Figure 2.16 could be implemented to reach study objectives faster with optimal test coverage.

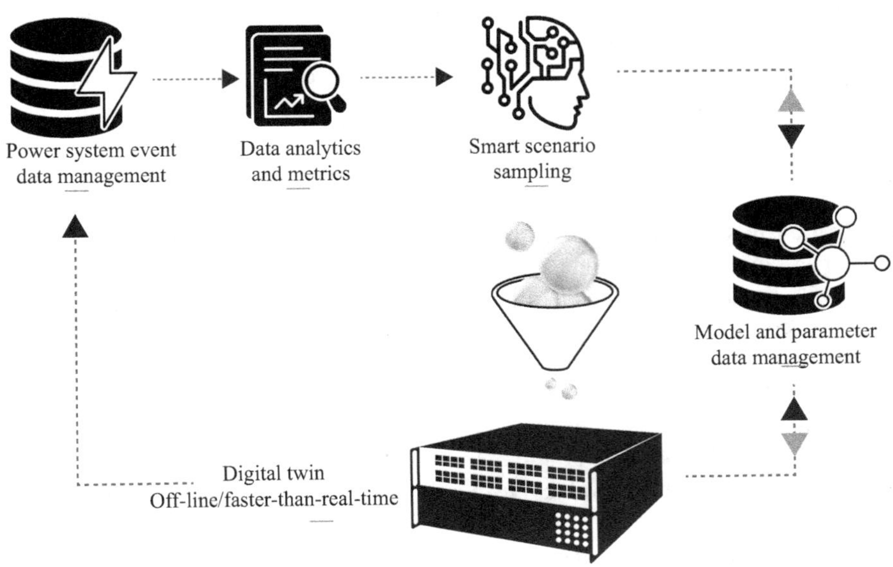

Figure 2.16 Automated AI-based optimal test scenario selection for contingency and event analysis

2.4 RTS testing of smart inverters

2.4.1 Smart inverters at the heart of the smart distribution system: control architecture in smart distribution systems

The distribution system is experiencing several technological evolutions and will continue to do so for the coming decades. This includes the integration of new power-electronics-based technologies and renewable energy sources with improved efficiency and flexibility, controllability, and varying impact on grid power quality. New technologies with faster power electronic switches and faster controls are emerging to provide higher efficiency, power density, reliability, and faster response. Distributed energy resources (DERs) and controllable loads are becoming more common, where the generation is distributed throughout the grid, as opposed to centralized upstream of a radial configuration, making it a challenge to control coordination and interaction. With colocated generation and loads, smaller micro-grid systems are becoming popular; these typically work synchronously with the rest of the distribution system but are able to isolate themselves intentionally or due to various grid contingencies and then operate autonomously. When distribution conditions become favorable, the microgrid can then reconnect to the main utility grid.

The characteristics of DERs in distribution systems, such as non-dispatchability of renewable-energy-based generation, bring about various new challenges to distribution system operators, microgrid operators, and independent power producers. One way to tackle these challenges is by managing DERs as a controlled unit and allowing the distribution systems and the microgrids to be isolated from the rest of the main utility grid and sustain themselves. This leads to newly arising concepts in distribution system control architecture (Mohammed *et al.*, 2019). As depicted in Figure 2.17, one widely known architecture involves three control levels: level 1, level 2, and level 3.

The third level is where the utility objectives are controlled, whether this is in terms of managing the power reserve from independent producers to managing power transfers or through providing economic dispatch. The large amount of prosumer (an energy consumer who also produces energy) equipment that will eventually be installed in the distribution system will require an increasing amount of AI-based decision-making support algorithms for the operators to efficiently dispatch resources to the distribution system infrastructure. It will be a challenge to ensure that set points provided by level 3 controls will not cause instabilities due to the violation of power transfer limits, or to power imbalance among the load, the generation, and the voltage variation. Therefore, system operation may need to be verified at the operator level by using RTS-based DTs of the power system, including as many controls as possible and through testing multiple scenarios.

The second level is where energy management controllers are found in various forms and contexts in the distribution systems, including microgrid control

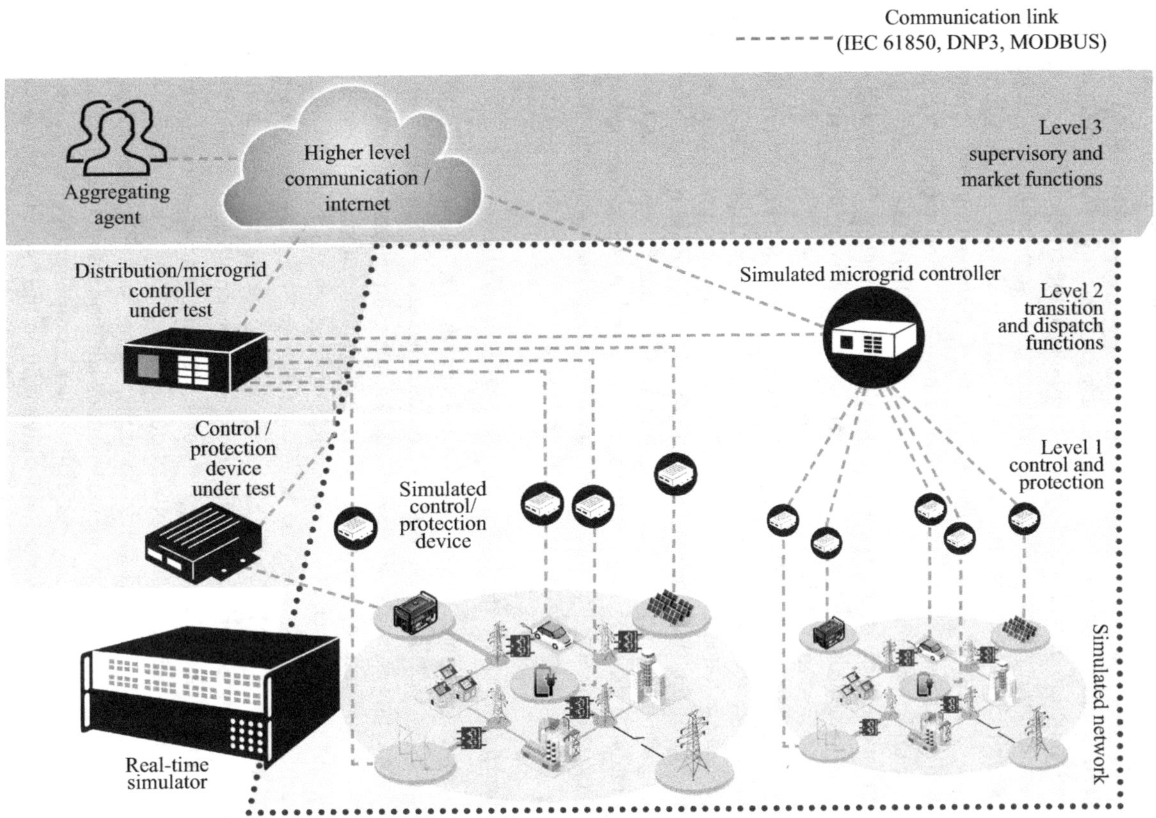

Figure 2.17 Overview of microgrid and distribution system architecture with focus on HIL studies

systems, DER management systems, advanced distribution management systems (ADMS), and EV charging station management controllers, to name a few.

Energy management controllers, as the "brain" of the integrated distribution system, are often responsible for power dispatch of DERs and for shedding/controlling the loads in different operating modes, as well as for the transition between grid-connected and islanded modes for specific sections of the distribution system, within a limited vicinity of a microgrid or a client of the distribution system owner. The dispatch function provides appropriate set points to the controllable grid assets, (DER, loads) either predetermined by the operator or by some sophisticated control algorithms subject to grid requests and constraints, load, and weather forecasts. These types of controls come in various forms, including (but not limited to) multiagent decentralized architecture (Han *et al.*, 2018) and more deterministic centralized rule-based control algorithms (Sun *et al.*, 2020). The use of CHIL to validate the operation of these types of controllers is increasing in demand, especially as new standards emerge with the requirement of testing their dispatch and transition functions thoroughly (IEEE, 2018b).

Real-time simulation and HIL can be useful in all control levels, but in the following sections, we mainly focus on the HIL testing techniques in the context of smart inverters, which are found at the first level of the control architecture. These inverters are at the heart of the smart distribution system, as they provide many control and protection functions for grid support, ensuring voltage and frequency stability and proper operation of the microgrid and the distribution grid. It is therefore important to test them first, making sure that their basic functionalities perform as intended and that they do not create adverse interactions with the rest of the distribution system.

2.4.2 Smart inverter design and testing using HIL

PV systems require power converters that convert the DC output from the solar cells to the AC so that it can be used for common electrical equipment and loads. Such DC-to-AC power electronic devices are known as inverters. PV inverters can be classified depending on the types of AC connection, functionalities, inverter topologies, and sizes (Farret, 2013). Regardless of these classifications, typically all commercially available PV inverters have the following components: power electronic devices (e.g., IGBT, MOSFET, and diodes), magnetics (e.g., inductors, capacitors, and transformers), controllers (e.g., microcontrollers, DSP, and FPGA), thermal and mechanical elements (e.g., heat sinks, fans, and enclosures), sensors (e.g., current and voltage), and protection elements (e.g., contactors, circuit breakers, fuses, and voltage relays). A controller is the "brain" for the PV inverter that regulates the power conversion, incorporates additional functionalities, applies protection logic, and communicates with other components or systems, both internal and external to the inverter. When we refer to the "smart" qualities of the PV inverter, this typically lies in the control logic that determines the advanced functionalities it can provide when connected to the utility grid. Even though such an inverter can also operate in islanded mode without grid connection and even capable of transitioning between grid-connected to islanded modes for microgrid applications, in the next few subsections, we will only discuss grid-connected operation. Additionally, it is important to note that smart inverters are currently more prevalent in the installed PV systems and battery energy storage systems. But they may also be utilized for other DERs such as wind, microturbine, fuel cells, and flywheel. Without going into various types of power conversion details for DERs as discussed in the study of Chakraborty (2013) to explain HIL applications, we mainly focus on PV smart inverters for this section.

The first generation of PV inverter controls was developed for PV systems not connected to the grid. These applications were primarily for remote locations where utility lines were either not available or not reliable. As the PV system cost decreased, residential and commercial PV applications started to increase, most of which required the inverters to connect to the utility grid and supply the loads in conjunction with the utility. Such PV inverters initially were only regarded as the active power source, running at the unity power factor. Also based on the IEEE Std 1547-2003 (IEEE, 2003) and other grid codes, they were usually prohibited from regulating the voltage at the PCC and needed to trip as soon as possible for grid disturbances. Due to the comparatively lower penetration of PV systems in the grid during these times, such basic inverter operations were considered good enough for the stability of the grid. As PV penetration has increased considerably in the last decade, inverter controls have also needed to provide more advanced functions such as voltage and frequency ride-through, voltage regulation, frequency response, and communication for monitoring and control as laid out in revised versions of the standards, such as IEEE Std 1547-2018 (IEEE, 2018a). This new generation of advanced inverters is often termed "smart inverters."

As controls for smart inverters became increasingly complex, it became imperative to have a better development, verification, and validation tool for design cycle of these inverters. Real-time simulation and HIL tools and techniques are beneficial for such applications and have extensively been used by both research communities and manufacturers. Furthermore, the HIL-based methods provide a way to test the power system level response of these smart inverters for large PV penetration scenarios and to test complex interactions between smart inverters themselves and between the smart inverters and other grid assets, such as voltage regulators, capacitor banks, and protection devices. In the next sections, we will discuss examples and applications of the use of real-time simulation and HIL for smart inverter development and testing.

2.4.3 Smart inverter control development using rapid control prototyping (RCP)

In smart inverters, controls are typically implemented in the controller board(s) inside the inverter. Such controller boards consist of processing units (such as microcontrollers, DSP, and FPGA) along with signal conditioning and inputs/outputs. The control algorithms are then implemented in the controller and tested for their operation. As the development of such controller boards themselves is challenging, it is often beneficial to decouple the hardware development efforts from the algorithm development process. Additionally, programming these controllers often requires software skills and expert knowledge for complex coding and integration. In such cases, the RTS can be used to develop and test the advanced control algorithms before they are implemented in the final control platform.

A typical RTS consists of processing power provided by standard CPU cores and FPGA chips along with many fast digital and analog input/outputs. Additionally, the RTS and processes provide precise time synchronization needed for controllers. Moreover, some of the real-time software provides for the development of control codes using graphical, model-based languages such as Simulink® and Xilinx® System Generator toolboxes rather than working with low-level languages such as C, C++, and VHDL. All these hardware and software characteristics of the RTS make it eminently suitable for control prototyping of the smart inverter. These types of efforts are often termed as RCP, as the development, risk, times, and cost are often much less compared to the traditional controller programming methods.

An illustration of RCP for the smart inverter is shown in Figure 2.5. In this case, the simulator is used to implement the lowest level switching control logic for the smart inverter. The output from the simulator is the pulse width modulation (PWM) switching signals that are then sent to the actual inverter gate driver/switches using digital outputs. The sensor measurements from the inverter are sent back to the simulated controller through analog inputs to close the control loop and test the closed-loop control algorithms. RCP has typically been used for traditional controls as utilized in the smart inverters available in the market. But a similar approach can be used to develop and validate the neural network or "fuzzy-logic"-based controllers or hybrid "neuro-fuzzy" controllers in the future.

2.4.4 Smart inverter control validation using controller HIL (CHIL)

CHIL is the most widespread application in which an RTS is used for smart inverter development and validation. In CHIL testing, the actual inverter controller hardware is connected to a real-time simulation of the converter models that can then be connected to a grid model. The control algorithms are implemented in the controller hardware and tested with the real-time model in a closed loop, causing the controller to "think" that it is connected to actual devices, whereas in reality, it is only connected to a simulated inverter and the rest of the power system model.

Over the years, the complexity of controls has significantly increased for PV inverters. Current smart inverters commonly have multiple layers of controls. At the lower level, closed-loop current control, synchronization functions, and protection functions are often implemented in the controller. For smart inverters with multiple conversion stages, typically a non-isolated or isolated DC-to-DC converter front-end connects the PV panels to the DC-to-AC inverter stage. For these topologies, the lower level controller also includes closed-loop controls for the front-end converter. These lower level controls usually operate at high speeds (from 10 to 100 s of microseconds) and produce switching pulses to the gate drivers for power electronic switches. The higher level of controls in smart inverters often produces reference signals (such as reference for the current controller) for the lowest level controllers based on the smart inverter power and voltage regulation functions and the operating modes.

Maximum peak power tracking, implemented either at the DC-to-DC or at the DC-to-AC stage, is also included in this control layer. Additionally, this higher level of controls can interact with supervisory controllers (such as a microgrid controller, and ADMS) or other external signals such as dispatch coming from a PV aggregator or an electric utility for providing grid services. The higher level of control is comparatively slow and often works within the timescale ranging from milliseconds to seconds. It is worth mentioning that there is no distinct physical boundary between high-level and low-level controls, and that often a single controller may perform all these functions in a smart inverter.

Because of the inherent complexity, CHIL is an important tool in the smart inverter development process to eliminate any errors in the control algorithms, controller hardware, or programming at earlier stages of development. If these errors are found later in development, mitigating them can be costly and time-consuming. Additionally, running the actual inverter hardware with a faulty controller increases the risk of catastrophic hardware failure and safety issues during testing. CHIL can also be used to test various hypothetical scenarios in the early developmental stages, such as fault response and islanding detection, without the need for costly testing hardware or a connection to a real-world utility grid (Prabakar *et al.*, 2017).

Depending on the timescale of the control actions, EMT-type smart inverter models can be simulated in CPU cores or in the FPGA of the RTS, or both the CPU and FPGA may be used. For example, testing the lowest level controls and protection functions of smart inverters, switching at tens of kilohertz or higher, requires models in the FPGA with a time step of 1 μs or lower. On the other hand, testing the higher level functions may require only simulation time steps in the range of 10–100 μs, if using EMT simulation with average inverter models, or in the order of 1 ms time steps if using TS simulation and can be simulated in CPU cores. An example of CHIL test setup for the smart inverter is shown in Figure 2.18.

In this setup, the detailed EMT-type inverter model with switches, filters, contactors, and sensors is modeled in the FPGA of the simulator and connected to the actual controller hardware. The PWM switching signals from the controller are sent to the simulated inverter model using digital inputs of the simulator. The outputs from the inverter model such as current and voltage measurements and contactor status are sent to the controller through analog and digital output channels.

In this example, a feeder model is simulated in the CPU cores and that model interacts with the inverter controller using analog and digital channels as well. The PCC voltage and current measurements from the feeder model is sent to the controller's higher level controls to execute frequency/voltage regulation functions and to generate reference signals for the lower level controls.

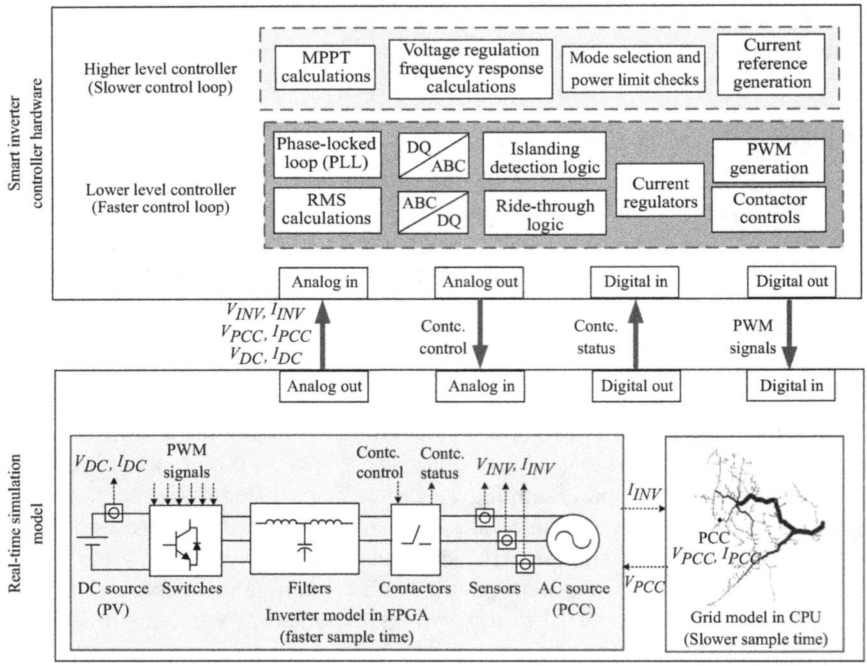

Figure 2.18 Example of CHIL for a smart inverter

2.4.5 Smart-inverter-power-system-level validation using Power HIL (PHIL)

PHIL simulation is another evolving application relevant to smart inverters. This type of setup requires a power amplifier that amplifies low voltage/current signals from the simulator to the real-world voltage/current signals so that the inverter can then be connected to the power amplifier and operated at full power. The measurements from the hardware setup are collected through sensors and fed back to the simulator to complete the loop. The details of PHIL testing and considerations related to amplifier selection, interfacing, etc. are discussed in the early part of this chapter.

As mentioned previously, the new generation of smart inverters includes various grid support functions such as ride-through, voltage regulation, and frequency response. As the PV systems are non-dispatchable, power outputs from them vary considerably over time depending on the solar irradiance. In combination with the daily and seasonal load variability, the variation of the PV in a distribution feeder with large amounts of PV can create conditions such as reverse power flows and voltage fluctuations. Furthermore, depending on the level of penetration, some of the impacts of PV systems can also be felt at sub-transmission levels. Similar problems exist for other DERs as well, especially when they are connected to the customers' side of the meters without direct control from the utilities. Therefore, the goal for these new smart inverter functions is to actively measure and respond to various grid conditions to provide stability and operational support to the grid. Consequently, it is becoming extremely important to test these smart inverters' interactions with the grid in a closed-loop fashion before they are deployed in the field (Lundstrom *et al.*, 2016).

In an ideal scenario, pure off-line simulation could be used for such power-system-level validation of the smart inverters if (i) all commercial inverters behaved exactly the same or (ii) if detailed models were available from inverter manufacturers for various inverter types. But in the real world, neither of these two conditions is the case currently. The inverters' response to faults and other abnormal grid conditions are often dependent on the lower level control implementations which can vary between manufacturers and models and are not available publicly due to intellectual property concerns. Testing the inverter using traditional test procedures, as done during certification, cannot be relied on to properly evaluate the operation of the inverter functions that are grid interactive such as voltage regulation (volt-VAR, volt-Watt) or frequency response (frequency-Watt).

Moreover, as the number of smart PV inverters from vendors using different control implementations rises, the likelihood of harmful interactions between their autonomous grid support functions also increases. Most of the work on detecting such interactions has been based on pure off-line simulation or analytical methods. Such pure simulation or analytical methods can be very effective, if the inverter models are properly known and the control details are validated through actual testing. However, in a traditional test setup, such testing is very difficult to conduct with multiple inverters, due to the need for multiple controllable voltage sources to represent PCCs with variable grid impedances between the PCCs.

CHIL, as discussed earlier, can be used in the initial development stages of the smart inverter to address these closed-loop testing challenges. In the later stages of product development, when the complete smart inverter hardware is available, PHIL provides a way to test such complex interactions. As with CHIL testing, PHIL testing can be used to validate control modes and inverters' responses to normal and abnormal grid conditions. But unlike CHIL, which validates only the controller, PHIL testing involves the actual inverter hardware, including power electronics, magnetics, sensors, protection devices, and cables/conductors that are parts of the interconnection. In this way, more non-idealities of real-world smart inverter systems can be included in the testing (Lundstrom *et al.*, 2016). The PHIL testing could also be used by utilities willing to homologate or validate commercial smart inverters' compliance with the interconnection requirements before they are installed in the field or to request improvements from the manufacturers. To determine the closed-loop interactions between the smart inverters and the grid, the grid model can be implemented in the RTS, and the power amplifier is used to connect the smart inverter hardware operating at full power as part of this simulated power systems model. The voltage and current measurements from the inverter can then be fed back to the simulated grid.

A simple PHIL test setup example is shown in Figure 2.19. In this setup, the real-time model consists of the simple Thevenin equivalent model of the grid PCC

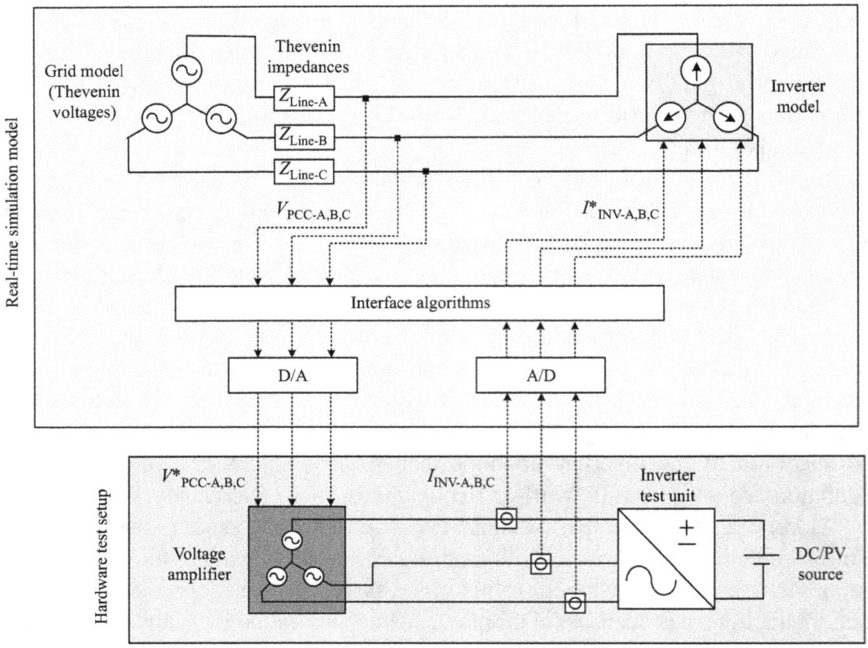

Figure 2.19 Smart inverter PHIL test setup with simplified grid model

to which the smart inverter is connected (Hoke *et al.*, 2015). The smart inverter itself is represented by the controlled current source in the real-time model. The interfacing between this model and the hardware also requires interface algorithms and compensation methods not depicted in the figure. The real-time model sends the voltage set points for the amplifier, and the amplifier generates the voltages based on these set points. The smart inverter is then connected to the amplifier and sends power to the amplifier. In this case, it is assumed that the amplifier is bidirectional and can accept the amount of power generated by the inverter. For a unidirectional amplifier, a separate load bank is needed between the inverter and the amplifier to sink the power generated by the inverter. The output current measurements from the inverter are then fed back to the model and based on those measured currents; the injected currents at the PCC of the model are controlled to complete the loop.

Some of the early works on smart inverter PHIL used simplified models for the grid and the inverter as shown in Figure 2.19 (Hoke *et al.*, 2015). These were later extended to include detailed grid models (Nelson *et al.*, 2016) and multiple smart inverters connected at multiple PCCs (Chakraborty *et al.*, 2016; Hoke *et al.*, 2018). Depending on the case study, the real-time grid model can be the reduced order or the full-scale grid model in the EMT domain (Nelson *et al.*, 2016). For very large power system models, co-simulation methods can also be used to connect a large electromechanical transient domain model (typically used for TS analysis) with the

EMT model in real time, and the smart inverters can then be connected to the nodes of the EMT model (Pratt *et al.*, 2019). Such co-simulation methods can be beneficial to capture not only direct interactions between the smart inverter and the PCC but also the scenarios when such interactions are spread over a large number of other PCCs over the power systems in high-PV penetration scenarios.

Until now, PHIL-based smart inverter testing was largely used to answer research questions related to interactions between them and between the smart inverters and the grid. However, one of the initial efforts for inverter PHIL testing was recently recognized by the North American interconnection test standard, IEEE Std 1547.1-2020, as an alternate method for testing unintentional islanding detection of the grid-connected inverters for PV and other DERs (IEEE, 2020). For the smart inverter interconnections to the utility grid, detection of islanding scenarios during grid faults is crucial for the safety of both equipment and personnel. Traditional islanding detection tests require not only the AC test source but also a resonant RLC load bank with similar rating to the test inverter. Additionally, the RLC loads need fine-tuning capabilities for conducting the test as required by the standard. Such tests can be burdensome in terms of cost and availability of the equipment, especially for large smart inverters. To overcome this challenge, PHIL-based methods were developed by researchers over the years (Lundstrom *et al.*, 2013). In the recently published IEEE Std 1547.1-2020, such testing is being accepted as an alternate procedure for equipment certification (IEEE, 2020).

In Figure 2.20, the PHIL-based unintentional islanding setup is shown. In Figure 2.20(a), the traditional test circuit is shown followed by a PHIL-based test circuit in Figure 2.20(b). In the PHIL-based circuit, the AC test source, islanding switch, and the resonant RLC load bank are parts of the simulation, and the simulated system then interfaces with the inverter through the power amplifier. The details of the PHIL-based test, their advantages and limitations, and the comparison to the traditional testing can be found in Lundstrom *et al.* (2013).

Verification of the effectiveness of unintentional island detection functions can become challenging to multiple smart inverters with advanced grid support functions and with interleaved grid impedances. PHIL-based methods can be used to address such testing challenges. An example of PHIL test setup is shown in Figure 2.21 that was used for multi-inverter, islanding detection testing for three inverters (Hoke *et al.*, 2018).

In this PHIL setup, inverters can be connected to various parts of the grid model thus creating scenarios where they are connected both to the same transformer and to different transformers with distribution lines between them (representing a typical solar subdivision). The flexibility of the PHIL test setup therefore can be useful to compare and validate various operational scenarios with respect to the number of DERs in the island, the topology and impedances of the interconnecting island circuit, and the type and location of the load within the island circuit. The details of the PHIL testing and the results can be found in the study of Hoke *et al.* (2018). A similar testing structure has been used to address interactions between the smart inverters providing reactive power support (Chakraborty *et al.*, 2015, 2016).

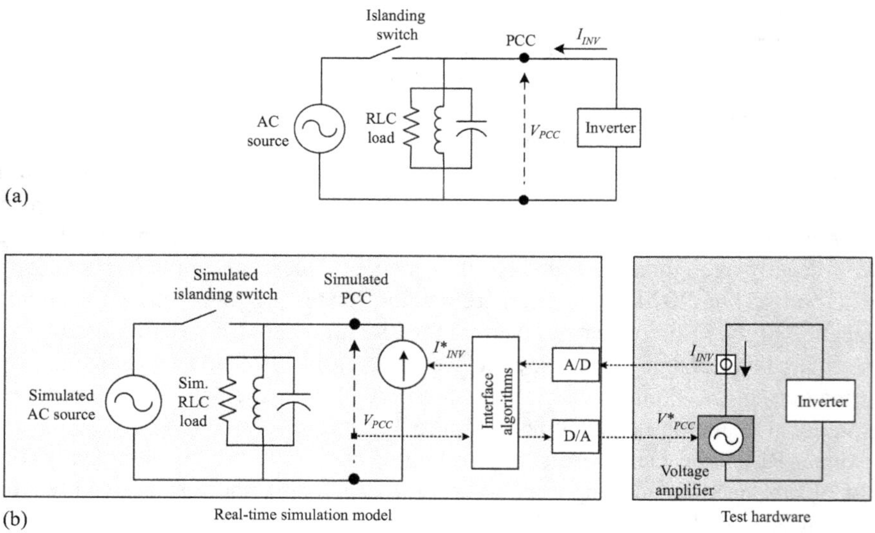

Figure 2.20 Unintentional islanding testing based on IEEE Std 1547.1-2020: (a) traditional test circuit, (b) PHIL-based test circuit

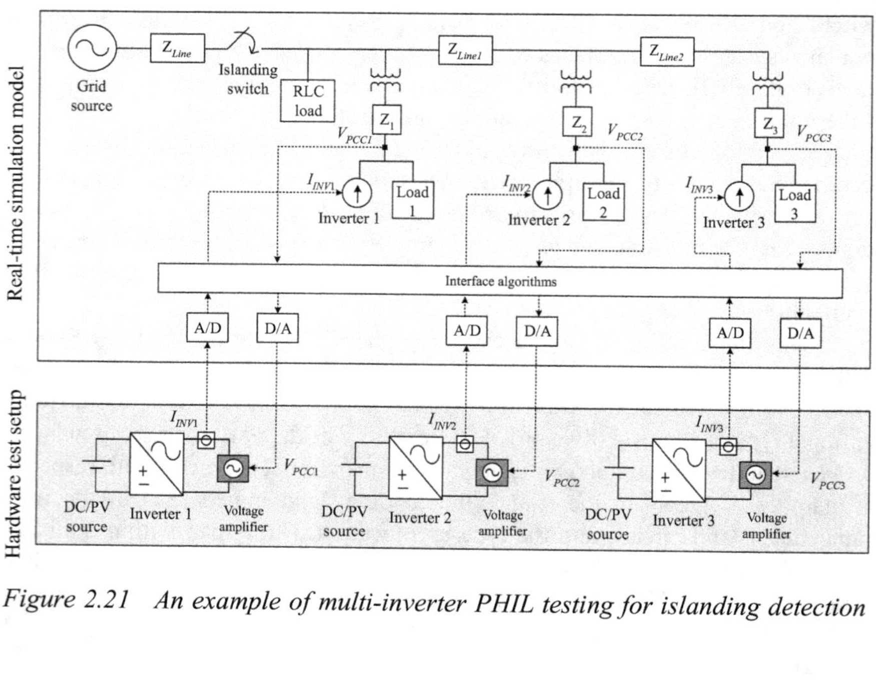

Figure 2.21 An example of multi-inverter PHIL testing for islanding detection

2.5 RTS testing of wide area monitoring, control, and protection systems (WAMPACS)

Before adopting synchronized measurement technologies (SMTs) in the power system industry, there was an interval between the moments when a local protection/control failed and when the operator in the control center could confidently react to an abnormal condition in the system. This gap arose from the fact that the conventional telecommunication protocols are able to transmit measurement and alarm data to the SCADA system every 20–30 s; additionally, the operator in the control center would need time to confirm that a variation in incoming data is due to a true failure in the system and not just a glitch or false alarm. Therefore, when that local/primary protection failed to respond, it took a minute or longer until an operator initiated a corrective or preventive action.

Wide area monitoring, protection, and control (WAMPAC) is a concept that fills some technological gaps in power system automation. As described in Terzija *et al.* (2011), "WAMPAC *involves the use of system-wide information and the communication of selected local information to a remote location to counteract the propagation of large disturbances*." SMT is the key component of WAMPAC systems. Exploiting PMUs as the most popular SMT-based devices in the power industry opened the door to collecting, transmitting, and archiving the time-stamped data significantly faster than with conventional telemetry technologies. The first prototype of the PMU was developed and tested at the Virginia Polytechnic Institute and State University in the early 1980s. The first commercial unit, the Macrodyne 1690, was developed in 1991. In the late 1990s, the Bonneville Power Administration developed a wide area measurement system, initiating the use of PMUs for large-scale power systems (Chakrabarti *et al.*, 2009).

Although PMU (and its variations such as micro-synchrophasors; Von Meier *et al.*, 2014) is the enabler of WAMPAC systems, there are other crucial components in such a system. Global positioning systems are required by PMU to time stamp each measured signal with a precision of microseconds; a phasor data concentrator (PDC) is required to collect time-stamped measurement signals from widely distanced PMUs and to produce time-aligned output data streams; a telecommunication infrastructure is required to cement these pieces together and to transmit the data to various applications.

Three major wide-area applications are involved in such systems: monitoring and analysis, control, and adaptive protection. For each application, several types of functionality have been deployed (Martinez *et al.*, 2005) as follows:

- frequency data collection;
- frequency response and rate of change monitoring;
- voltage angle/magnitude and VAR monitoring;
- power flow corridors;
- state estimation;
- open phase detection;

- disturbance/oscillation monitoring;
- pattern recognition;
- spectral analysis, mostly for electromechanical oscillation modes;
- model validation;
- online thermal rating of transmission lines;
- system instability (transient, voltage, and frequency);
- power swings; and
- out-of-step detection.

Figure 2.22 (inspired by Terzija *et al.* (2011)) shows the block diagram of possible applications that might be used in an integrated WAMPAC system in different control layers. Despite differences in objectives of these applications and their capabilities, their common goal is to boost power transfer in transmission and distribution systems while maintaining system reliability. Since these novel technologies are still under development, it is of paramount importance to test them during challenging operational scenarios before considering system operating decisions based on their output. Additionally, system operators must be well trained with each new technology to both trust and feel at ease with using it during stressful contingencies. Digital RTSs are the most suitable platforms for addressing such pre-implementation concerns.

Several pilot projects and programs are running around the world with the aim of implementing and installing WAMPAC systems to improve or resolve the issues that the conventional control and protection systems are unable to address. Begovic *et al.* (2005) and Gavrilas (2009) examine some examples of WAMPAC system developments in various countries in 2005 and 2016, respectively. Two of these referenced projects are revisited here.

The first case study is from TNB, an electric utility company in Malaysia, where a real-time application platform (RTAP) has been developed. This platform is designed to collect data from multiple sources with various communication protocols, to process these data, and to execute control commands to multiple controllers. RTAP is targeted to be used for smart grid applications, in centralized or distributed architectures, and to facilitate control and monitoring of substations or IEDs. However, before deploying their RTAP in an operational system, it was tested extensively through RTSs. Figure 2.23 shows the real-time simulation setup used to verify the operation of the RTAP in identifying and controlling the transient instability in the Malaysian transmission system (Sarmin *et al.*, 2018).

In this setup, the RTAP acquires required measurements through PMU streams (C37.118) in real time from the simulator (labeled as RTPSS); then it performs the real-time analytical operations deployed in its processor, and issues control signals through the IEC 61850 protocol, finally sending them back to the simulator. Choosing the right RTS for such an application is an important factor that can significantly influence both the overall cost of the project and the quality of the testing. Most of the control and protection functions related to WAMPAC applications required to be tested and verified can be covered by TS simulations rather than by detailed electromagnetic simulations. Additionally, as a WAMPAC system

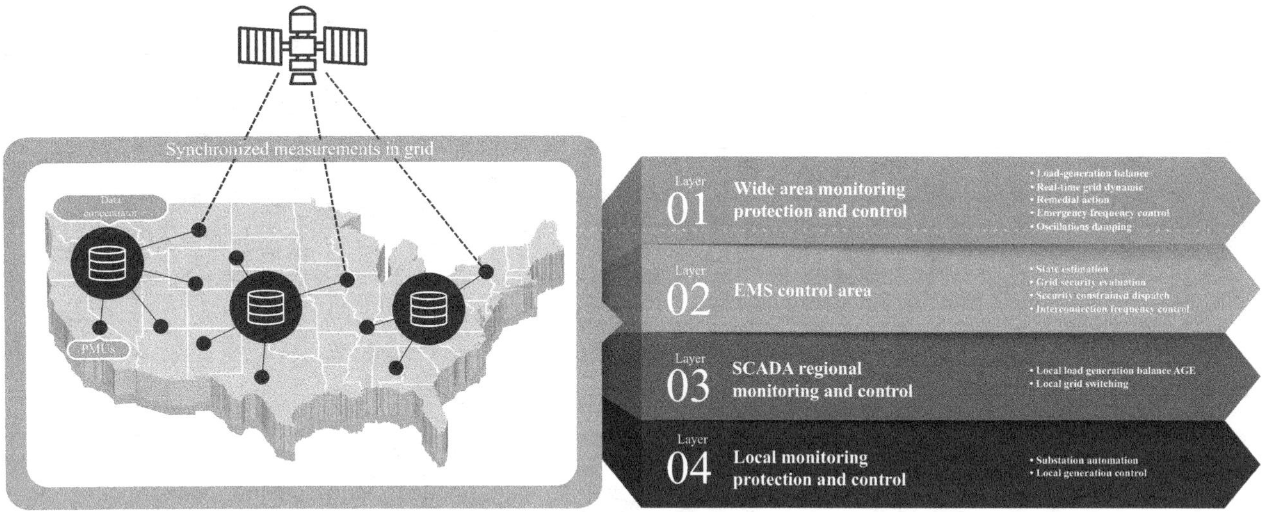

Figure 2.22 Different WAMPAC applications for each layer of power system

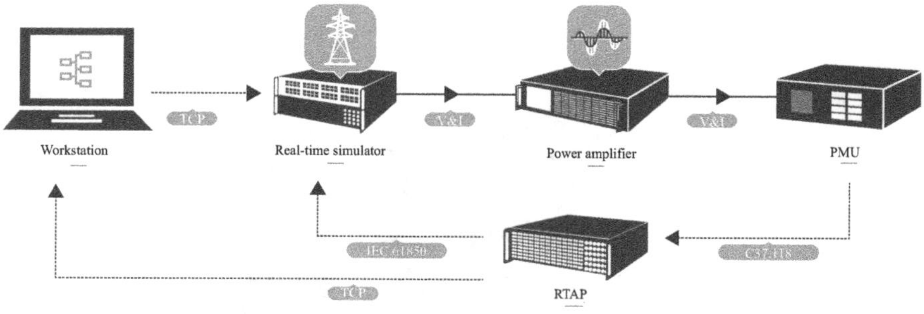

Figure 2.23 RTAP HIL testing platform

implies (through its name) that it will deal with large-scale grids, especially when the scope of testing involves both transmission and distribution systems, the size of the system can increase to thousands of buses. Considering this, the engineers at TNB have chosen OPAL-RT's ePHASORSIM (Jalili-Marandi *et al.*, 2012, 2013) toolbox as the real-time TS simulator to build up their suitable test platform (Azmi *et al.*, 2019).

The other case study is from IREQ, Hydro-Québec's Research Institute in Quebec, Canada, where they have initiated a project, called global and local control of compensators, aiming at deploying a WAMPAC system to maintain voltage stability via controlling several shunt compensators installed in the network. In Quebec, there is a long distance (1,000 km) between principal energy resources (hydro generation located in the north of province) and major energy consumers (large cities in the south). HQ network—the sole power utility in the province—uses twelve 735 kV transmission lines that connect the resources to the loads. Several types of reactive power compensators have been installed in the network to control and improve the voltage stability throughout the network using local voltage measurements. However, researchers at IREQ found that for the future evolution of the power system, in the case of a severe contingency resulting in a voltage drop in the southern part of the grid, the total available reactive power capacity of all installed compensators would not be used to keep the system stable and secure. This was a concern for system planning and operation.

Among the various candidate solutions for the future system, the most cost-effective one was to implement a WAMPAC system. The concept is to continuously monitor voltage in sensitive spots of the network (in this case, Montreal, as the single biggest load in the system), and to transmit that measurement to geographically dispersed substations. In the case of voltage drop detection, the local controller regulates the operational set point for shunt compensators to contribute to reactive power injection to the system to avoid voltage collapse. After proof of concept was done using off-line simulation tools, to verify the accuracy and robustness of the proposed WAMPAC system, it was connected to a replica of the HQ network modeled in HYPERSIM, the RTS from IREQ. This replica includes detailed modeling of substations and all the equipment installed in the current HQ system. Therefore, several contingency scenarios can be duplicated in the replica, by pushing the limits to discover the worst cases, and through execution of the HIL tests, the sanity of the WAMPAC system's operation can be validated. These tests convinced the researchers and management to go one step further and to install the WAMPAC system and its infrastructures in the real grid (Ghahremani *et al.*, 2019).

In addition to the smart inverters and WAMPAC systems, as discussed in this chapter, the current trend in the power system industry is evolving toward having more complex systems for control and operation in the future, and thus more vulnerability to failures. Increasing numbers of nonutility generators, independent energy producers, and high levels of power interchanges, along with constraints imposed by economic and environmental parameters, are just some of the factors that have made power utilities worldwide feel the need to find smarter solutions to identify, predict, decide, and act as fast as possible when an emergency occurs in the power network. Fuzzy logic and neural network paradigms are promising in this regard; however, they require enormous datasets to train.

2.6 Digital twin concepts and real-time simulators

2.6.1 RTS-based digital twins: DT background, key requirements and use cases

RTS technologies are key technologies to deploying DTs (Grieves, 2016), either on conventional control center computing platforms or through using high-performance computing in the cloud (Kagermann and Wahlster, n.d.) for use in future power systems. Based on the concept of HIL and the available interfacing media in RTS technologies, a connection between the virtual (simulated) world and the physical hardware can be built. The foundational principle of digital twinning, in this context, is that the virtual and the physical worlds can interact with one another in real time.

Before delving into specifics, first of all, a survey of the appropriate background knowledge regarding digital twinning and its development throughout the year must be carried out. Figure 2.24 records the developing milestones of the DT in history. The timeline goes back to 1970. A digital mirror system was initially proposed by NASA to investigate the possible cause of explosion during the Apollo 13 mission in a simulation environment (Barricelli *et al.*, 2019). The idea was to simulate and monitor the states of spacecraft in mission. This concept then evolved into product lifecycle management (Caruso *et al.*, 2010) in 2002. In this version of the concept, the key requirements were clarified for the first time. Three requirements were presented: a real physical system, a virtual model, and dataflow between the physical system and the virtual one, linking the two systems together. This link remains throughout the entire lifecycle of the real system for the data exchange.

The term "digital twin" was introduced 10 years later in the aerospace realm by Eric Tuegel (Tuegel *et al.*, 2011). The airframe DT was proposed in order to build

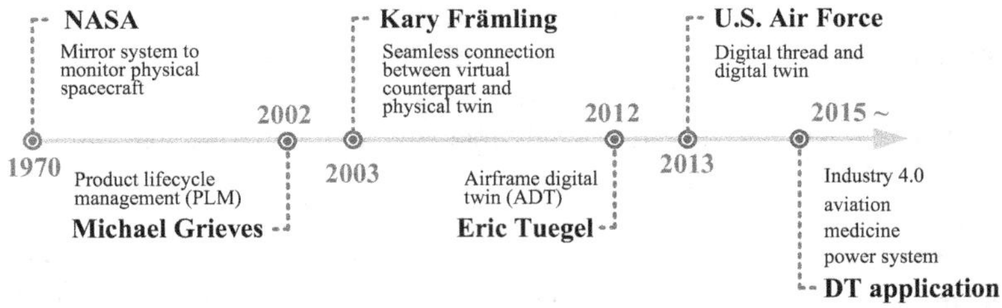

Figure 2.24 Timeline of digital twin (Barricelli et al., 2019)

the computational aircraft model to predict and schedule the maintenance for the physical aircraft. One year later, the U.S. Air Force (2019) explicitly mentioned the digital thread and DT concepts, emphasizing that they had the ability to exploit the previous and then the current knowledge to monitor the states and predictively diagnose the system, thus providing the adaptability needed for rapid developing. Since the first declaration of the DT concept by the USAF, it has received much more attention beyond the aerospace realm. In Industry 4.0 and Smart Manufacturing (Barricelli *et al.*, 2019), the DT concept has become an essential part of allowing digital manufacturing and cyber-physical systems to develop from 2015 onward.

The successful application of the DT in aerospace has motivated research and implementation to some extent in manufacturing and industrial fields. In these areas, DT is defined as a computer-based digital model or a set of digital models that can mirror its physical counterpart that receives information from the physical system, accumulates useful information, and helps in decision-making and in the execution of processes. DT in the manufacturing arena uses computer-based digital models to monitor the procedure in production process, and with the assistance of AI algorithms, an autonomous and intelligent manufacturing approach is executed with minimized human intervention. It can respond to failures or unexpected contingencies with automated decision-making drawn from a set of alternate actions to prevent damage to the whole process at the supervisory level. Also, the connectivity between a DT and one or more physical systems allows current and historical data analysis by human experts assisted by AI algorithms to derive solutions toward improving operations (Mechatronic Futures, 2016).

Regarding the DT concept as outlined here (OPAL-RT TECHNOLOGIES, 2020), the RTS provides further implementation of it in the power system field. In power system applications, a virtual model is intended to be a dynamic, evolving, and even an "intelligent" entity so that it changes over time as the physical system evolves (e.g., even with regards to physical parameters of system components).

To understand how the RTS works with the DT concept, let us first consider three key attributes of the DT:

1. a digital model in a simulated environment,

2. the physical entity in real space, and
3. a connection between the virtual model and the physical entity for the dataflow.

Here the RTS platform is a key component in the twinning process. A simple understanding of the twinning process arises from replicating the dynamic states from the physical system in the digital environment. This duplication process is adaptive and occurs in real time. Any slight change, however, can be detected and mirrored in the DT. Figure 2.25 depicts the twinning process and the DT utilizing RTS technologies. It implies data measurement and data collection (here in the form of synchrophasor data, but it is not limited to this) from the physical power grid. Normally, the hardware to be tested is directly connected with the I/O interface in an RTS. The simulation of the digital model runs in real time and it has its data communication through the I/O interface. In this figure, the hardware in power systems is not directly connected to the simulator, but it links the measurement database in the power system with the media interface such as the communication protocol.

Before we can understand realization of the DT concept in power systems through HIL concepts, we need to understand two types of DT concept: DT **prototypes** and DT **instances** (Grieves, 2016).

A DT **prototype** is used to portray the prototypical physical product. It contains sets of necessary information to describe and generate a physical component that duplicates the virtual model. There is no link between the DT and the physical system in this type. The informational sets are utilized to build the initial virtual model. A DT **instance**, on the other hand, contains the link between the DT and a physical product throughout its lifecycle. It describes the operational states of the corresponding physical products in real time. With this basic knowledge, we first need to build the DT prototype, according to the topology of the physical power grid. The measurement equipment like PMU measures the operational states of the power grid, which includes the state amplitude and angle (θ) regarding the

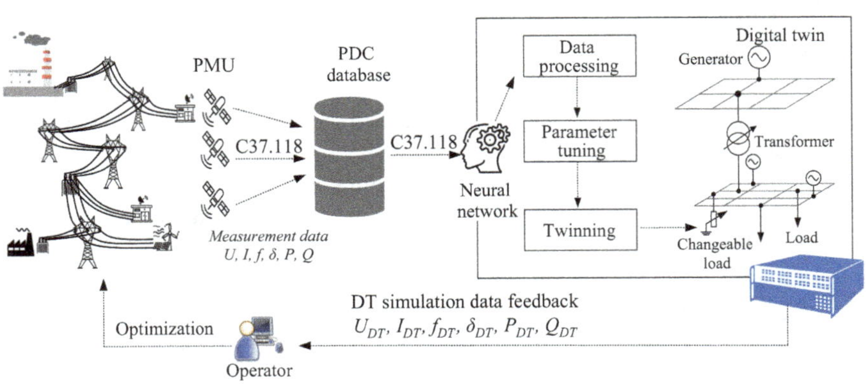

Figure 2.25 DT in hardware-in-the-loop simulation

fundamental component of voltage U and current I, commonly in the positive sequence, but certain PMUs can monitor individual phase synchrophasors. The fundamental frequency and rate of change of frequency of the power grid can also be measured by PMUs. The active and reactive power (P and Q) can be calculated and then all the measured data are transferred to the PDC database through the communication protocol, e.g., C37.118. With the media interface of the RTS, the database is accessible to the RTS. To build a DT for a large-scale power grid, the transmitted data to digital simulator is enormous, including the operational state of the individual components such as generators, overhead lines, and transformers. With neural networks, the operational data of these components are first identified and then classified for the parameter tuning process.

Before the DT prototype evolves into a DT instance, except for the link between the PDC database and the digital model, parameter tuning is a critical procedure. Due to the individual components, it is necessary to set up one or more different parameter tuning algorithms. The transmitted big data is classified into the categories, according to the requirements of these algorithms. These data are later utilized for parameter tuning, ensuring more accurate dynamics of the DT. The mirroring process is completed by continuously adapting the parameters of the digital model that dynamically duplicates the operational state of the physical power grid. The DT can simulate the potential events to support the operator for testing of the measures like the protection actions and new concepts. Based on this test, the operator can set up effective measures to optimize the physical power system.

The DT has many applications such as predictive simulation of events like fault scenario analysis and contingencies, to help support decision-making in power system operations. An instance of the DT could be used for off-line planning and design studies, e.g., to revise the control and protection system philosophy or the adaption in the case of network expansion. Other applications are considered next, such as predictive maintenance, supervision, diagnostics, and optimization of operations to support the decisions of the operator in power control center.

2.6.2 Model parameter tuning and adaptivity

Parameter tuning is an essential step before the digital model evolves from DT prototype to DT instance. With the supervised learning and pattern recognition algorithm (Wärmefjord *et al.*, 2017), the operation data and the surrounding IoT environment from the physical system are first identified and classified by a neural network. Then, these data are delivered to a tuning algorithm for parameter tuning. The parameter tuning can guarantee the adaptivity with the updating of the proper parameters in the model for replica of the physical operation state. Figure 2.26 outlines the complete process of parameter tuning and keeping the adaptivity of the DT to the physical power grid active.

The operation state in the physical power grid is first measured by the measurement equipment, e.g., PMU, which is transferred to the database via data communication as with the communication protocol. After classification and

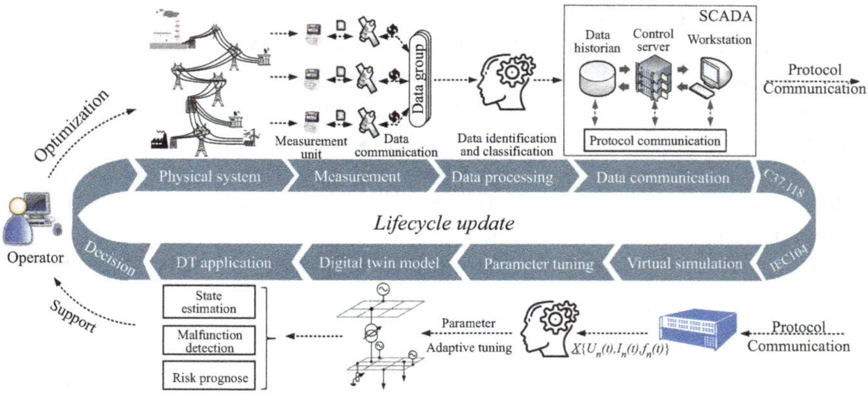

Figure 2.26 DT parameter tuning and adaptivity in lifecycle

identification of these data, it is stored in the database server of SCADA. Usually, the virtual model is not directly accessible to the database server for modeling the DT. The RTS provides the link to connect the virtual simulation platform with the database server via power system communication protocols like C37.118 and IEC 60870-5-104, or even IEC 61850 communication protocols. The parameters of the components such as generators, transformers, and loads are set up in the digital model with a parameter set from the simulator (from the initial set of parameters). The values of this starting set of parameters typically differ slightly from the real, measured data due to climatic conditions of the physical system or changes in the operating behavior of the components due to wear, etc. The adaptive parameter tuning process aims to solve this problem. By updating the physical operation data in the tuning algorithm, the parameter is tuned correspondingly and updated in the DT to keep the simulation as close to the real world as possible.

The application of the DT regarding its adaptivity can be utilized to estimate the states, the malfunction detection, and the prediction of potential risk for physical power grid (Song *et al.*, 2020). These applications support the further decision for operators to optimize the physical power grid for safe and effective operation. The cycle in Figure 2.26 shows the process of how the DT is always updating its parameters or features with the accompanying changes of the physical system. This lifecycle updating feature guarantees the adaptivity of the DT for further applications.

2.6.3 Cyber-physical surveillance and security assessment

The cyber-physical system exploits the available digital data of the physical power system and its environment. RTS technologies allow continuous data exchange between the cyber and the physical world. The DT is characterized by its ability to monitor the physical system accurately and adaptively on different scales of time. Accordingly, the DT can be a part of the cyber-physical system, which interacts with the physical entities, e.g., equipment, environment, and humans.

The physical power grid communicates with virtual cyberspace through HIL simulation. Each component in the physical power grid has its digital representation. Since the DT replicates the real world as closely as possible, the system can also be monitored on the basis of the DT. Figure 2.27 outlines the interaction between the real network, the DT, and the applications based on it. The DT is thus the data supplier for applications such as stability considerations, forecasts, and condition monitoring. In this sense, the DT takes on the current role of an SCADA database with subsequent state estimation.

This provides the operator with information about the system status and the development of the system state. The information base is much closer to the physical process than it could be with other processes. It is even foreseeable that in the future, the operator will receive proposals for action for network operations based on this information. However, the algorithms required for this are still the subject of basic research. There is a high probability that a large area of application for machine learning will develop here.

2.6.4 Predictive simulation and operator decision support

With DT, predictive simulation is possible to predict the operating status and possible critical events in the physical power grid. In the event of a fault event such as a short circuit in the nearby generator, the system can use the DT's predictive simulation to determine how this will affect the stability of the entire system. This approach is useful for evaluating and optimizing the operation by preemptively informing operators of the required maintenance in advance of the expected failures. Basically, this is a DT-based DSA network. As shown in Figure 2.27, the operator can use the predictive simulation results to support decisions in the control room. Using AI algorithms such as machine learning, the data can be extracted with feature information to detect changes and to identify key patterns and trends.

Figure 2.28 outlines how DT applications and DSA interact with the control room GUI to support the operator's decision. The SCADA database has access to the PDC database to receive the measured physical data from the PMU. The remote terminal unit, a measurement and communication unit in a substation, provides the

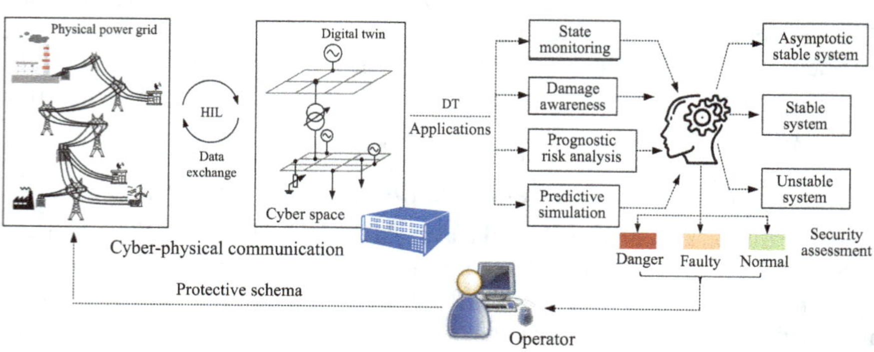

Figure 2.27 Cyber-physical surveillance and security assessment

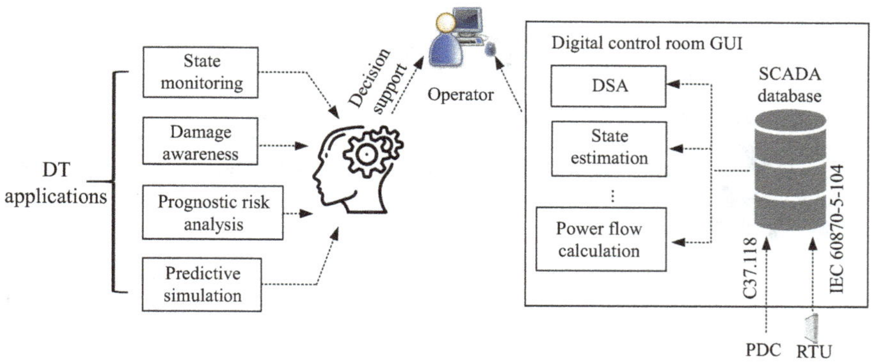

Figure 2.28 Decision support by DT application

measurement data to the SCADA database via the IEC 60870-5-104 protocol or other communication protocols. These data are used in DSA, status estimation, and power flow calculation in the digital control room to determine the operating status in the physical power network and to make it available to the operator for evaluation. In the future, the DT will provide additional predictive analysis to estimate future states and important changes in the power network to further support the operator in optimizing operations.

2.6.5 Detecting equipment malfunction

The DT always runs parallel to the operation to monitor the operating status of the real power grid. If there is a large deviation between the real measured data and the states of the DT, there is an indication of a malfunction in the real system. The operator can be warned immediately if the operating behavior deviates from the expected predictive simulation or the DT simulation.

The data of the deviation characteristics are identified by AI algorithms such as machine learning to initiate the appropriate repairs. The entire detection process is divided into two steps. First, in the AI algorithm development phase, the developer defines the corresponding DT model, and according to the exchanged data, the DT is correctly parameterized with the characteristic features from the physical power grid (see Figure 2.29). Possible disturbance events are then examined using the DT instance to generate data patterns with which typical events in the real world can be identified.

The second phase is the training process and validation. Machine learning is used to form a detection model by learning the fault simulation data in the first phase. It should be noted that the training data must be sufficiently good both qualitatively and quantitatively to ensure that the event detection is as selective as possible. The objective is to combine the DT with the AI algorithm to perform this task. This procedure can not only accurately predict and identify malfunctions in the early phase of the actual operating state in the power system but also quickly prepare the appropriate reactions such as maintenance measures to prevent damage to the power system.

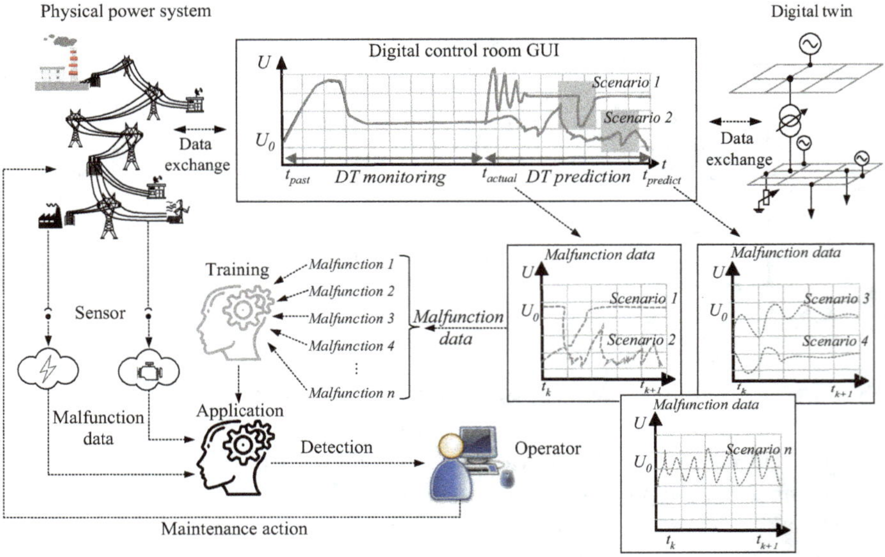

Figure 2.29 Malfunction detection with DT

2.6.6 RTS as a key enabler toward implementing AI-based digital twins and control systems

From the outlined explanations and applications of DT, it should be obvious that the capability of simulating the complete power grid in real time (or by utilizing the RTS architecture and capabilities to accelerate the simulations) is required and of clear benefit while providing monitoring, parameter tunings, and scenario analysis to support system operators. Furthermore, the introduction of distributed generations using complex and fast power electronic systems may require the use of detailed EMT simulation to capture the performance of future power systems, which will react faster than conventional power grids do. Consequently, fast EMT RTS with the capability of simulating very large grids will play an important role in enabling the use of AI-based DTs and control systems.

Contributors to this chapter

Jean-Nicolas Paquin, OPAL-RT Technologies, Canada
Jean Bélanger, OPAL-RT Technologies, Canada
Sudipta Chakraborty, OPAL-RT Technologies, USA
Vahid Jalili-Marandi, OPAL-RT Technologies, Canada
Xinya Song, TU Ilmenau, Germany
Dirk Westermann, TU Ilmenau, Germany
Marcelo G. Simoes, MGS Engineering, USA

Chapter 3
Fuzzy sets

During the twentieth century, particularly after the advent of computers and advances in mathematical control theory, many attempts were made for augmenting the intelligence of computer software with further capabilities of logic, models of. Adaptive learning algorithms were developed, making possible the initial developments in neural networks in the 1950s. A very innovative learning approach was birthed by L. Zadeh in 1965 with the publication of his paper "Fuzzy Sets." In that paper, the idea of a membership function based on multivalued logic, allowed a solid theory where technology bundled together thinking, vagueness, and imprecision. An engineering design starts from the process of thinking, i.e., a mental creation, and designers will use their linguistic formulation, with their own analysis and logical statements about their ideas. Then, vagueness and imprecision are considered as empirical knowledge to be incorporate in the model implementation of the system. Scientists and engineers try to remove most of the vagueness and imprecision of the world by making precise mathematical formulation of laws of physics, chemistry, and the nature in general. Sometimes, it is possible to have precise mathematical models, with strong constraints on non-idealities, parameter variation, and nonlinear behavior. However, if the system becomes more complex, the lack of ability to measure or to evaluate features, with a lack of definition of precise modeling, in addition to many other uncertainties and incorporation of human expertise, makes almost impossible to explore such a very precise model for a complex real-life system. Fuzzy logic (FL) and NNs became the foundation for the newly advanced twenty-first century of smart control, smart modeling, intelligent behavior, and artificial intelligence (AI). This chapter discusses the basics and foundations of FL and NNs, with some applications in the area of energy systems, power electronics, power systems, and power quality.

3.1 What is an intelligent system

Control systems make possible a response to a given input in accordance with their transfer function; intelligent systems are the ones capable of supplying answers to solve problems; especially fitting specific situations but also capable to deal with new or unexpected circumstances. Intelligent systems approach unique solutions, creative ones, designed to mimic nature and biological systems; for example: (i) observing how a person implements some predefined control functions, and (ii) looking for patterns on data or behavior, and taking decisions on the basis of historical experience. Although a lot of achievements in the past decades demonstrated a great computational power of computers and software capable of learning and doing outstanding modeling and analysis, most of hype are present in science fiction movies. There is still a gap in how humans think and act in a creative way when compared to how computational machines implement their decision-making. In such a perspective, a person is capable of holding two opposite concepts in their mind and still come up with an attitude that might be completely rational and unexpected. People may think in uncertain ways, with imprecise data, and with facts that are blurred, whereas computers will be moved by an algorithm written in a precise and mostly binary way, i.e., having a workflow defined by yes/no paths and true/false statement evaluations. When a human will make a decision if a baked treat is good, bad, or wonderful, the evaluation will be made in what could be considered uncertain, imprecise, or what has been defined in the past few decades as a "fuzzy way."

AI is a discipline for studying how people solve problems and how machines (computers) may emulate such human behavior on "problem solving," in other words, how to make machines to have further and deeper attributes of human intelligence.

Fuzzy Logic (FL) is a technique to incorporate the human nature in thinking in a control system; a typical FL controller (FLC) could be designed to behave in accordance with the deductive reasoning, i.e., the process in which people use to infer conclusions on the basis of information known with previous experience. For example, human operators can control industrial processes and complex manufacturing plants, which could even have nonlinear mathematical models and not completely defined dynamics, based on experience, inference, training with more experienced tutors. FL can capture such knowledge in a fuzzy controller, allowing the computational implementation of an algorithm that has equivalent performance of the human operator.

Another possible action of thinking would be logical and sensible when applying the inductive reasoning. The approach is to learn and generalize from unique examples fed by data and observation of dynamic process behavior, with time-varying conditions, in order to design a fuzzy controller. In such an implementation, the fuzzy system is taught, and the fuzzy controller adapts to a given performance, i.e., an adaptive fuzzy control system will learn from experience, when tasks are performed in a repetitive way, and a management layer will make the fuzzy controller to adapt and improve, based on a performance index or optimization function. Therefore, learning-by-example associated with encoding human expertise makes fuzzy systems very robust, extensible to being applied in a wide variety of engineering systems.

Controllers or regulators combining conventional and intelligent techniques are often utilized in the closed-loop control of dynamic complex systems, such as integrated and distributed power electronics for smart-grid-enabled power systems. Operational or supervisory fuzzy controllers consider a global strategy for management. The strategy could be either a supervisory control management, for example, in complex industrial process, or a supervisory energy management, such as a large and dispersed distribution electrical power system.

Operational tasks are usually delegated to people who might look at several synoptic panels with operator/system communication for process control. Supervisory industrial control systems would have experts looking into several set points in a process plan, observing how raw materials would be transformed by several machines and processes, with varying temperatures, flow, pressures, and fine-tuning PID controllers on the fly as the process never stops. For such complicated industrial manufacturing processes that can never be stopped, the experience of a human operator can be captured in a fuzzy controller, providing a heuristic approach for implementing supervisory algorithms in a computational environment. Similarly, the utility power grid can never stop, and decisions must be taken in accordance with load demand and generation availability, constrained to maximum loading of transmission lines, losses, heating, and substation's capacity. When a local generator starts or stops, or when electric plug-in vehicles are connected to a certain feeder, a supervisory intelligent controller can be implemented either by deductive reasoning, inferring conclusions based on information from experts, or by inductive reasoning, where repetitive behavior can be improved by data storage for adaptive learning of controllers to achieve a given performance. FL control is well fit for this type of application.

Another powerful tool in implementing intelligent control is the application of artificial NNs (ANNs), inasmuch they have the capability of learning how to provide classifications, or data estimation, or control actions based on numerical data associating input and output, whereas fuzzy control works better with semantic examples.

An intelligent control may allow the design of an autonomous system, i.e., those that could execute complex control tasks under all operating conditions for a process, or a system, resilient to faults, without supervision or interventions of external operators. Several space control missions have budget and resources to design such autonomous systems. In the past few years, we have been experiencing the development of autonomous driving cars, enhanced factory automation, and many other applications. However, total autonomous intelligent systems, with creative capacity are yet to be developed for the current technological generation.

Some questions that could be thought as a reflection on the subject of this chapter are as follows:

- What is the difference between conventional and intelligent systems?
- Why an FL control would be considered as intelligent?
- What is the main reason to study, understand, and design intelligent systems?
- What would be the main characteristics of an FL-based control versus an ANN-based control?
- What are the main characteristics and features of an intelligent control?
- What is the main difference in designing a control system based on deductive reasoning versus inductive reasoning?
- Where exactly intelligent control systems improve and enhance power-electronics-enabled power systems, smart-grid systems, and integration of renewable energy systems?

You are invited to think about these questions, and how enhanced engineering design could be applied to advance electrical systems, and improvements in power-electronics-enabled power systems.

3.2 Fuzzy reasoning

FL remained a theoretical concept with little practical applications for a couple of years after being introduced by Prof. Lotfi Zadeh in the 1960s. Zadeh proposed initially mathematical ideas and philosophical reasoning. In the 1970s, Prof. Ebrahim Mamdani of Queen Mary College, London, built what is considered to be the first practical fuzzy system for a steam engine controller. Later, there was an application of fuzzy systems for traffic light control. With those practical applications, there was an extensive development of fuzzy control applications, and many products became embedded with FL. Japanese scientists started to use the theory of FL into many technological advances during the 1970s. It is anecdotal history that FL was not well accepted in English-speaking countries due to stereotyping preconception that a logic could never be fuzzy, while in other non-English-speaking countries, such discrimination was not ever factored by people who decided to use that technique. The initial successful fuzzy set control devices were manufactured in Japan, using analog control and later microprocessor-based implementations. By the end of the twentieth century, there were many FLCs implemented in a plethora of products, from washing machines to speedboats, from air condition units to handheld autofocus cameras, plus industrial automation and power system transmission and distribution control. The first Ph.D. dissertation in applying both FL and NNs for power-electronics-based renewable energy management control of grid connected wind turbine systems was defended and published in 1995 (Simões, 1995).

FLCs have ability to cope with knowledge represented in a linguistic form instead of the conventional mathematical framework. Although control system engineers have traditionally relied on mathematical models for their designs, it has been observed that as more complex becomes a system, the less effective is their mathematical model. Real-world problems can be very complex. Therefore, FLCs can be useful in incorporating human experience, intuition, and heuristics into the system controllers, instead of relying on in-depth mathematical models. FL control can be very effective in applications where existing models are ill-defined and not reliable.

In a classic set theory formulation, given a universe of discourse, U, one can define the collection of objects that may belong to this universe of discourse. The classical set is a collection of a number of those elements where they are either a member or a nonmember element of the set, so their membership is either 0% or 100%; in Figure 3.1(a), there is a Boolean logic representation clearly indicating that the element x belongs to a set, while elements y and z do not belong to that set, all of them are in a domain, or universe of discourse U. On the other hand, Figure 3.1(b) shows some shading around x to illustrate that the element x has a

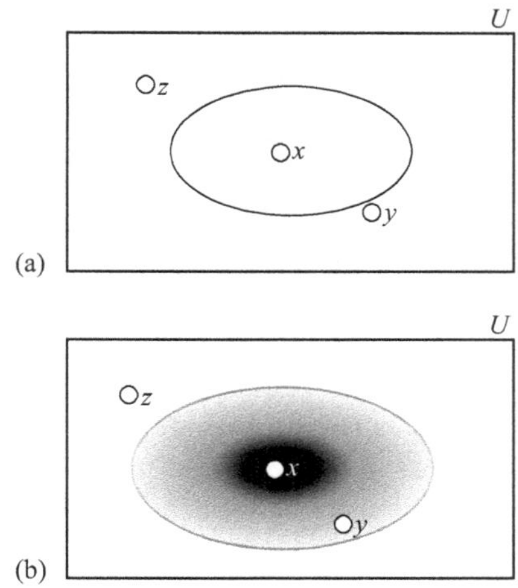

Figure 3.1 Boolean and fuzzy sets: (a) classical/crisp set boundary and (b) fuzzy set boundary

degree of belongingness or pertinence, while element z is illustrated in the way to show that it does not belong at all and there is also a gray pertinence of element y because it has some gray area that makes y to also belong to a certain degree as x to such a fuzzy set. As such, Figure 3.1(a) shows typical crispy, or Boolean, sets, where an element indeed belongs or does not belong, while Figure 3.1(b) illustrates shades of gray, where there is a partial membership of elements. If you look at the grass in your home and you ask if "is this grass green?" you may say yes or no, and try to define in a Boolean way. However, if you look carefully you can see grass that is sharp green, light green, yellowish green, and orangish green, and you can define that there is a degree of membership of the quality of the grass being green, and you can accept fuzziness for such a real-life observation.

A crispy set could be denoted by $A = \{x \epsilon U | \mathrm{P}(x)\}$ where the elements of A have the property P, and U is the universe of discourse. We can describe a characteristic function $\mu_A(x) : U \longrightarrow \{0, 1\}$ defined to value "0" if x is not an element of A and "1" if x is an element of A. For this evaluation, U will encompass only two elements, "1" and "0." Therefore, an element x, in the universe of discourse, is either a member of set A or not. Such a precise Boolean property can be applied to some real world or maybe natural events, but it may create problems, or ambiguity. So, where would be such ambiguity about membership on a set? Let us discuss the vagueness of human nature descriptions next.

Let us illustrate ambiguity: consider the concept "bald man" for a male individual. Such a concept is clear in our mind. You and I would recognize a bald man. But how to precisely determine the beginning and ending of such a transformation of someone who has a "luxuriant flowing hair" to someone who will eventually become "bald"? This is called the "bald man paradox." Would you describe a man with one hair on his head as bald? Yes. Would you describe a man with two hairs on his head as bald? Yes.

If you keep continuing and doing such a reasoning one step at the time, it would be possible to arrive to the question: would you describe a man with 10,000 hairs on his head as bald? Mostly not. Therefore, in such sequential reasoning one can draw a line, splitting the concepts of (i) one who is bald versus (ii) one who is not bald? Such discussion is general and philosophical in nature. Let us have a renewable energy and power system of similar idea: consider the concept "dry season" for a couple of months in a year and it did not rain. In such a concept, a season is a natural period into which the year is divided by the equinoxes and solstices or atmospheric conditions. If it does not rain enough, water reservoirs would be "not full enough" for hydroelectric-powered turbines to properly generate electricity. Both farmers and electrical generation companies would prefer a rainy season instead of a "dry season." A draught would be an extreme situation as in that there was no rain at all. A dry season would have rained very little to maintain dams and water reservoirs full, the same sequential idea of how much should it rain to have a rainy season, to have a rainy season, under a fuzzy perspective: how rainy would be good enough for agriculture, recreation, and electrical power generation needs?

The same paradox can be applied in defining a middle-aged person, adult person, a pile of sand, and heap of grains; they are philosophically defined as "the sorites paradox." It is a type of paradox that arises from vague predicates. We may have a definition of the concept "middle-aged" implying that a person who is 34 years old would suddenly become middle-aged on his next birthday and would suddenly, just after the day of her 56th birthday, this person would no longer be middle-aged. As we could also think about an adult person who suddenly becomes adult after their 21st birthday, or maybe after their 18th birthday, or for car insurance policies, after his or her 26th birthday.

Let us take the notion of a "comfortable outdoor temperature" and ask people around the world which temperature is the one they would feel comfortable outside. We will get different values, from Middle East to Northern Europe, to Bahamas, Alaska, Patagonia, Siberia, Portugal, Brazil, Rocky Mountains or Japan, and probably at the beach on California it would be different than at the winery regions in California, the very same state. Predicates are the part of a sentence that expresses what is said about the subject; human beings have their own perception of their own descriptions, which might conflict with a precise mathematical description of their own environment. The scientific method application on observing the real world can be visualized in Figure 3.2. The world phenomena will have some data and analysis, which will give a model for the scientist or engineer; such a model might be based on equations, graphs, and diagrams. This understanding will lead a decision-making process to handle the variables on the real world, in order to have a constructive and operational functional design. In such a kind of thinking, under vagueness and imprecision, there is a notion supported in the "Principle of Incompatibility" by Lotfi Zadeh. Prof. Zadeh stated in his first published paper (Zadeh, 1965) plus in another article written a couple years later (Zadeh, 1973) on fuzzy sets: "As the complexity of a system increases, human ability to make precise and relevant (meaningful) statements about its behavior diminished until a threshold is reached beyond which the precision and the relevance become mutually

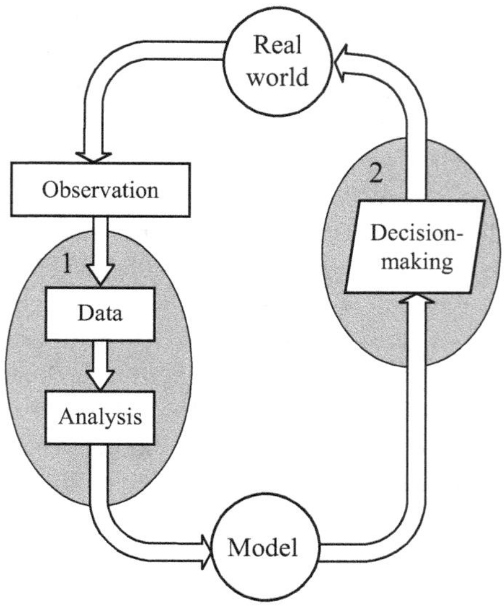

Figure 3.2 Scientific methodology observing the real world and performs analysis, in order to obtain a model for decision-making acting on variables that may control the real-world phenomena

exclusive characteristics so, the closer one looks at a real-world problem, the fuzzier becomes the solution." He also published in 1978 (Zadeh, 1978) about the possibility theory and soft data analysis. The principal constituents of soft computing (SC) are (i) FL, (ii) NN theory, and (iii) probabilistic reasoning; also subsuming belief networks, evolutionary computing, DNA computing, chaos theory, and learning theory.

Imprecision and complexity are correlated, i.e., when little complexity is present, closed-loop mathematical-based formulations are enough to describe systems. As more complex systems are under consideration, it may be required further AI-based solutions to reduce uncertainty. This leads to the discussion that when systems are complex enough, and only a few numerical data exist, and most of this information is vague, fuzzy reasoning can be used for manipulating such information. Fuzzy reasoning (set and logic) allows an SC methodology and implementation algorithm, with embedded intelligence, semi-unsupervised use of large quantities of complex data, uncertainty analysis, perception-based decision analysis and decision support systems for risk analysis and management, computing with words, computational theory of perception, and incorporation of natural language.

3.3 Introduction to fuzzy sets

We can define a set by using a function to define membership. For a typical Boolean or also called crispy set, such a function takes only two values, either 1 or

0, either True or False. However, for a fuzzy set, such function is continuously valued from 0 to 1, and associated with an element in the domain of the set; the range of such domain is called universe of discourse, i.e., $\mu_A(x)$: U [0, 1].

For such evaluation, an element, x, in the universe of discourse will have a membership value to such a set A defined by $\mu_A(x)$, which is defined as a fuzzy membership function for a fuzzy set A. When using classic Boolean logic, a set will be defined by a two-element characteristic function, either "1" or "0," or "True" or "False," and the element will "Belong" or "Do Not Belong" to the crispy set. Fuzzy sets theory extends the concept of sets to encompass vagueness. Membership to a set is no longer a matter of "true" or "false," "1" or "0," but a matter of degree. A typical comparison is to say that in a Boolean set, the membership is either Black or White. In conventional set theory, a scale of physical values could indicate a range of elements, for example, "0–100 V"; a value could be an element of this set (such as "25 V"), or it would not be an element of this set (such as "134 V"); on the other hand, in a fuzzy set, the membership has all shades of gray. The degree of membership becomes important that can take up any value between the unit interval [0, 1] and an element x will be partially, to a certain degree, a member of A, depending on the value of $\mu_A(x)$.

Figure 3.3 shows two membership functions, one is $\mu_A(x)$ for a set A, and the other is $\mu_B(x)$ for a set B, the variable x must be within the universe of discourse, i.e., in the domain or range $0 \leq x \leq 10$. When $x = 4$, the degree of membership for set A is 0.6, while the degree of membership for set B is 0.25. That means this variable belongs to both sets, to a matter of degree. The sets may have a semantical identification, or a name, or not been named at all.

The membership function $\mu(x)$ could be either a continuous or a discrete function. Although mathematical analysis would make better continuous cases, in digital computers, the set of sampled variables as well as tables allocating decisions requires discrete and finite values. Therefore, it is also important to consider that membership functions might be discrete. For example, Figure 3.4 illustrates the velocity of car, and assuming a threshold of 80 km/h, there is a bivalent association of speeders. However, such a sharp transition of speeding drivers would apply if

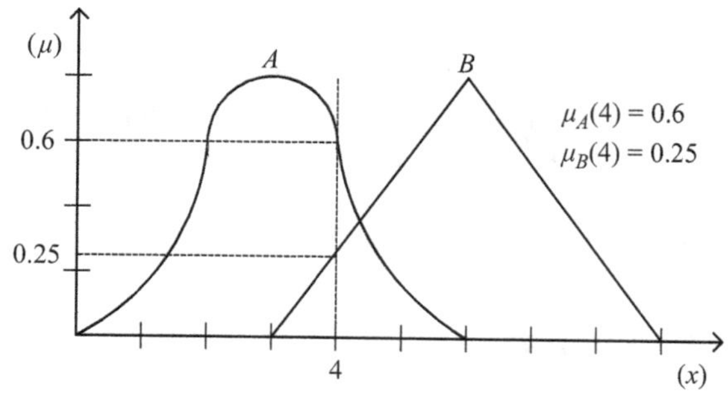

Figure 3.3 Membership function $\mu_A(x)$ for set A and membership function $\mu_B(x)$ for set B. The universe of discourse ranges $0 \leq x \leq 10$

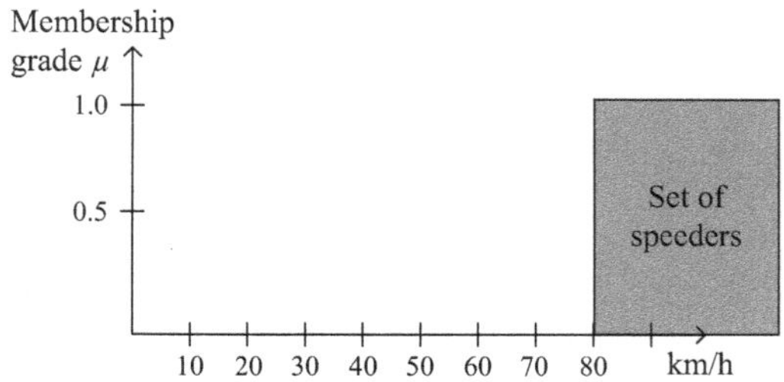

Figure 3.4 Example of drivers who violate the maximum speed limit, Boolean membership function

they were driving at 80.5 km/h, or maybe at 80.8 km/h. In practice, police officers know the imprecision of their measuring devices and would in their mind probably assume their own limit, say 83 km/h, or maybe 85 km/h, before taking their police car to chase the speeder for issuing a speeding ticket. Such a simple example clearly shows the mismatching of crispy set theory for practical aspects in life that allows multivalence, or ambivalence. Probably, a better membership function for such speeding situation analysis would be to allow

$$x_1 = 78.0 \ \mu_A(x_1) = 0.0$$
$$x_2 = 80.0 \ \mu_A(x_2) = 0.2$$
$$x_3 = 82.0 \ \mu_A(x_3) = 0.4$$
$$x_4 = 84.0 \ \mu_A(x_4) = 0.6$$
$$x_5 = 86.5 \ \mu_A(x_5) = 0.8$$
$$x_6 = 88.0 \ \mu_A(x_6) = 1.0$$

The behavior of such membership functions is depicted in Figure 3.5, where the transition of not-speeding to speeding is gradual, instead of sharp, maybe a local police officer would prefer to match the membership $\mu(x) = 1.0$ for drivers who strictly disobey the law (80.0 km/h), while in other driving zones, roads, or different localities, the police officer would assume that at 83.0 km/h the drivers would be at $\mu_A(x) = 0.5$, and could not yet be ticketed, depending on the evaluation.

Fuzzy sets theory is based on a real-life conundrum, where precise limits are not possible; a fuzzy set is a group of imprecise and not well-defined elements, where the transition of not-belonging to belonging to that group is gradual and not abrupt. A fuzzy characteristic implies uncertainty and qualitative definitions. Fuzzy sets theory provides a methodology in manipulating such human nature perspective. The uncertainty of an element is a fraction of their degree of pertinence to the set, and therefore, the concept of possibility becomes different from probability. If there is a green leaf, the probability (also in the range of 0–1) of finding a green leaf on the ground does not indicate if nature of green color is very green, blushing green, or greenish brown, but a fuzzy set would define such possibility of a green leaf found on the ground.

A very similar discussion could be considered in evaluating the membership function of a voltage measurement in power distribution feeder. Probably, a

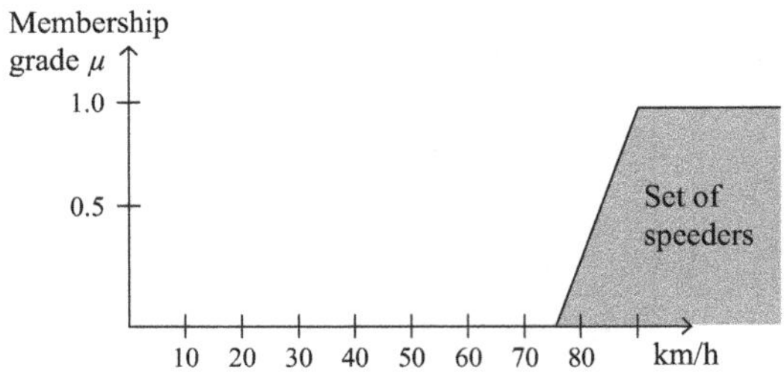

Figure 3.5 Example of drivers who violate the maximum speed limit, fuzzy membership function

protection relay could have a sharp and abrupt transition for a condition of "normal voltage" to "overvoltage" tripping and isolating a segment of the feeder, causing an interruption of electric power to some customers. However, in a renewable energy, smart-grid, power contemporary system, maybe some PV panels would be feeding power to the grid, elevating the voltage profile in the feeder, and certainly such a relay would have to be "further intelligent" and "with more data," in order to command an interruption of electrical power, and fuzzy sets plus FL could be the approach in taking the real-life considerations of a feeder protection relay in the case of deep renewable energy penetration.

3.4 Introduction to fuzzy logic

A linguistic variable would be a label associated with a fuzzy set, where such a fuzzy set is an ordinated pair of elements associated with their membership function. Therefore, the membership function becomes also labeled with the related linguistic variable. For example, a car velocity could be defined, linguistically, or semantically, as $T(velocity) = \{low, medium, fast\}$, on a *universe of discourse* such as $U = [0, 100]$, i.e., a range from 0 km/h to 100 km/h, where *low*, *medium*, and *fast* would terms, labels, or linguistic identifications of the variable *velocity*. Three membership functions could be defined along the horizontal axis representing the universe of discourse: maybe the variable *low* would have a left shoulder and a decreasing value from a maximum toward zero; *medium* would start from zero, having an increasing value, an apex or plateau, with a decreasing value (a convex shape); and *fast* would have an increasing value rising toward a right shoulder. These three fuzzy sets with corresponding membership functions would have proper overlapping in those functions. Therefore, the variable velocity would intersect two membership functions, making it possible to be at the same time, to a certain degree, for example, low and medium, as well as medium and fast. When there are fuzzy sets on the same universe of discourse, it is possible to apply at least three operations: NOT, AND, and OR.

Boolean set operations such as complement, union, and intersection are straightforward definitions in classical set theory. However, in fuzzy sets theory, those operations must be conducted with the membership functions. Zadeh (1965) proposed the fuzzy set operation definitions as an extension to classical operations. Although there are other algebraic formulations for defining the resulting membership function for union and intersection, often the following ones are used, due to their simplicity:

- **Complement (Not)** $\forall x \in X : \mu_{A'}(x) = 1 - \mu_A(x)$
- **Union** $\forall x \in X, \forall y \in Y : \mu_{A \cup B}(z) = \boldsymbol{max}[\mu_A(x), \mu_B(y)]$, where $z \in Z$, and all X, Y, and Z share the same universe of discourse.
- **Intersection** $\forall x \in X, \forall y \in Y : \mu_{A \cap B}(z) = \boldsymbol{min}[\mu_A(x), \mu_B(y)]$, where $z \in Z$, and all X, Y, and Z share the same universe of discourse.

These definitions form the foundations of the basics of fuzzy sets theory. The relationship between an element in the universe of discourse and a fuzzy set is defined by their membership function. The exact nature of the relation depends on the shape or the type of membership function used. Logic operators such as complement, union, and intersection are used when the variables have the same and common universe of discourse. The operator *complement* departs from Boolean logic, since an element can be partially assigned to a degree of true for a certain fuzzy set. For example, if we define linguistic variables for ages such as OLD and then NOT OLD, this does not mean completely YOUNG, and NOT YOUNG will not mean entirely OLD.

3.4.1 Defining fuzzy sets in practical applications

There are several methodologies to define fuzzy sets in practice (Zadeh, 1989):

1. **Polling:** The question "Do you agree that x is A?" is stated to different individuals.
   i. An average is taken to construct the membership function. Answers are typically yes/no type.
2. **Direct rating:** "How A is x?" This approach supposes that the fuzziness arises from individual subjective vagueness.
   i. The person is made to classify an object over and over again, until it becomes easier for him or her to remember the past answers.
   ii. The membership function is constructed by estimating the density function.
3. **Reverse rating:** The question "identify x which is A to the degree $\mu_A(x)$" is stated to an individual or a group of individuals.
   i. Responses are recorded and normal distribution for each are formed (mean and variance are estimated).
4. **Interval estimation:** The person is asked to give an interval that describes best access to x. This is suited to random set-view of membership functions.
   i. Membership exemplification: "To what degree x is A?" A person may be asked to draw a membership function that best describes A.

5. **Pairwise comparison:** "Which is a better example of A, x_1 or x_2, and by how much?"
   i. The results of comparisons could be used to fill a matrix of relative weights and the membership function is found by taking the components of the eigenvector corresponding to the maximum eigenvalue.
6. **Clustering methods:** Membership functions are constructed by a given set of data.
   i. Euclidean norm is used to form clusters on data.
7. **Genetic algorithm techniques:** Evolutionary approach is done to optimize the cluster of data into fuzzy sets.
8. **Neuro-fuzzy techniques:** NNs are used to construct membership functions.
   i. An essential part of forming membership functions is the input space partitioning. A grid that is fixed beforehand does not change later. It is set to some "initial value" and tuned.
   ii. Fuzzy clusters are best suited for classification problems, because they implement a similarity measure.

3.5 Fuzzy sets kernel

There are many mathematical definitions for intersection (AND) and union (OR), as indicated in Table 3.1. In order to establish a complete system of FL, one needs to define all these logic operations:

- Intersection (AND)
- Union (OR)
- Complement (NOT)
- Empty
- Equal
- Containment
- Alpha-cut

The literature on fuzzy sets presents and discusses many ways to calculate the fuzzy operation for AND (intersection) and OR (union). However, for most of the engineering modeling and analysis, two possible implementations, MIN/MAX and ALGEBRAIC, are done, as portrayed in Table 3.1, with a comparison of mathematical membership function definitions for OR (union) as well as AND (intersection). The

Table 3.1 Comparison of mathematical membership function definitions for OR (union) as well as AND (intersection)

	Union	**Intersection**
MIN/MAX	$MAX[\mu_A(x), \mu_B(x)]$	$MIN[\mu_A(x), \mu_B(x)]$
ALGEBRAIC	$\mu_A(x) + \mu_B(x) - \mu_A(x) \times \mu_B(x)$	$\mu_A(x) \times \mu_B(x)$

membership function for the intersection $A \cap B$ is represented by a mathematical membership function, and it can be defined analytically or calculated pointwise, i.e., a Cartesian product, such as $\mu_{A \cap B}(x) = \mu_A(x)\, \boldsymbol{t}\, \mu_B(x) \leq MIN[\mu_A(x), \mu_B(x)]$, where $\boldsymbol{t}$ is called ***t-norm***. The operator ***t-norm*** is an algebraic mapping, where two input fuzzy membership functions will give an output membership function. Similarly, the membership function for the union $A \cup B$ is also represented by a mathematical membership function, and it can be defined analytically or as Cartesian product given as $\mu_{A \cup B}(x) = \mu_A(x)\, \boldsymbol{s}\, \mu_B(x) \leq MAX[\mu_A(x), \mu_B(x)]$, where $\boldsymbol{s}$ is called ***s-norm***. The operator ***s-norm*** is an algebraic mapping, where two input fuzzy membership functions will give an output membership function.

There are transformations on an individual fuzzy set, such as the operation **alpha-cut** that is a crisp set formed by elements *A*, membership grade of which might be *s* greater than or equal to a given value *alpha*, such operation is usually used for decision-making. When it is necessary to refine descriptions, making them more meaningful and accurate, fuzzy sets can be modified, for example, to reflect with operators called hedges. A hedge can be applied to a fuzzy set, and the degrees of membership of the members of the set are altered. For example, Figure 3.6 shows a membership function for ***FAST***. The membership of ***FAST*** could be altered by ***VERY FAST*** by an exponential operator, as indicated in Figure 3.6. On the other hand, SOMEWHAT could also serve as a hedge to modify ***FAST***. The mathematical definitions are given by the following equations:

$$\mu_{VERY,A}(x) = [\mu_A(x)]^{\beta} \tag{3.1}$$

$$\mu_{SOMEWHAT,A}(x) = [\mu_A(x)]^{1/\beta} \tag{3.2}$$

The hedge "APPROXIMATELY" uses the concept of subsethood. It requires an auxiliary fuzzy set (distribution of weights) to evaluate assertions like "how much approximately cold is hot?" as indicated in Figure 3.7, i.e., it is a measure of how a membership function is correlated to another one.

Given a universe of discourse *E*, and three fuzzy sets $A \subset E$, $B \subset E$, and $C \subset E$, then the following applies:

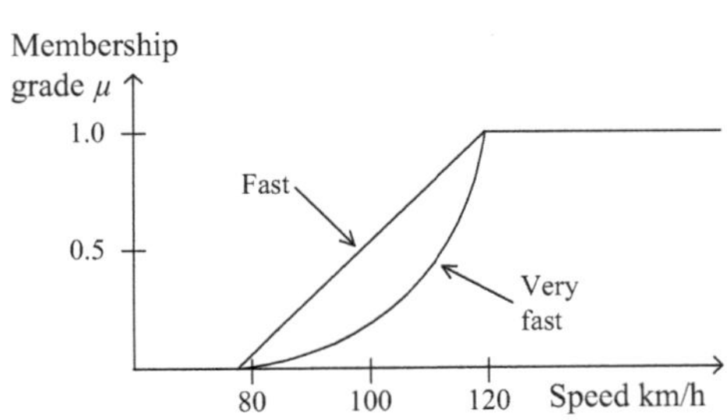

*Figure 3.6 Hedge function; fuzzy set **FAST** becomes **VERY FAST***

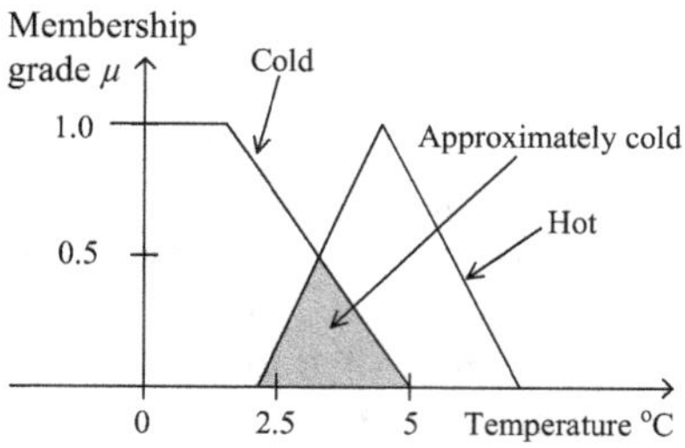

Figure 3.7 How much approximately cold is hot

Commutativity properties:
$A \cap B = B \cap A$
$A \cup B = B \cup A$
Associativity properties:
$(A \cap B) \cap C = A \cap (B \cap C)$
$(A \cup B) \cup C = A \cup (B \cup C)$
Idempotence:
$A \cap A = A$
$A \cup A = A$
Distributivity with respect to intersection:
$A \cap (B \cup C) = (A \cap B) \cup (A \cap C)$
Distributivity with respect to union:
$A \cup (B \cap C) = (A \cup B) \cap (A \cup C)$
Fuzzy set and their complement (*):
$A \cap A' \neq 0$
$A \cup A' \neq E$
Fuzzy set and the null set:
$A \cap \oslash = 0$
$A \cup \oslash = A$
Fuzzy set and the universal set:
$A \cap E = A$
$A \cup E = E$
Involution property:
$(A')' = A$
De Morgan's theorem: $(A \cap B)' = A' \cup B'$ $(A \cup B)' = A' \cap B'$

As an example, let us assume an air-conditioned vent. Suppose it has blades that control the openings and can be controllable so as the inclination angle of the vent might be directed downward or upward. Such angle control sends the airflow toward the floor or to the ceiling. Figure 3.8 shows fuzzy sets downward and upward, describing the position of the vent blades. If the blades are totally rotated to −45° in respect to the horizontal plane, then they are completely downward. If the blades are totally rotated to +45°, they are completely upward; the figure also shows the resulting membership functions when applying AND, and OR operations, with both fuzzy sets.

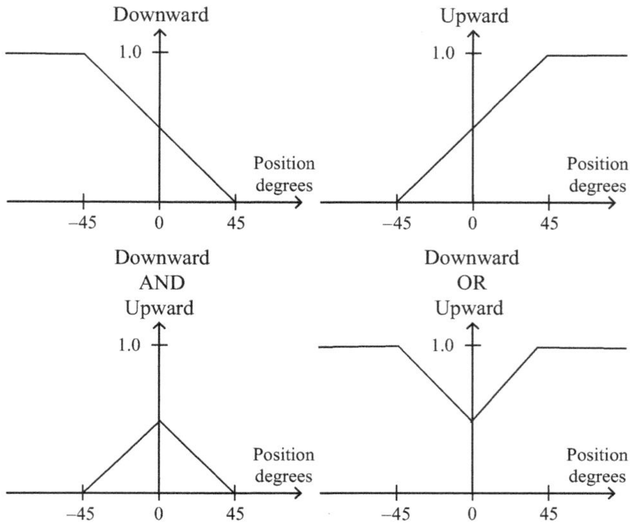

Figure 3.8 Membership functions for upward and downward, with corresponding AND and OR operations. The universe of discourse ranges $-45^\circ \leq \theta \leq 45^\circ$.

Membership functions can be defined with a graphical user interface (GUI), available in many simulation environments, such as MATLAB®, LabVIEW, as in several GUIs (GUIs for compiled or script languages, such as Python). In the past, i.e., from the 1990s to 2005s there has been several companies with visual-based development platforms, Togai, fuzzyTECH, and FIDE, just to name a few. Those GUIs, or "shells," provide real-time online debugging and tuning of rules, membership functions, and rule weights, including addition and deletion of rules, with a graphical object-based "point and click" and "drag" tools, with user-defined inference methods. It is also possible to have a fully integrated graphical simulation of fuzzy systems and conventional methods, with ANSI compiler, or another language. Those environments make provisions for reduction of programming efforts, fast prototyping, availability of several options regarding several degrees of freedom (typical of fuzzy systems), and visual feedback for assessing fuzzy controller modifications. There are several programmable logic controller (PLC) with fuzzy controllers and Fuzzy PID embedded on PLC programming, typically used for industrial complex systems, which may need human control operators for operation and supervision.

Chapter 4
Fuzzy inference: rule based and relational approaches

A fuzzy modeling system or a fuzzy controller can be implemented by understanding sampled valued variables, translated from the real-life world into a fuzzy domain. The model or the controller will estimate an output, either for a decision-mkaing action or to impose a set-point in a closed-loop control system. Those processes are defined as "fuzzification," i.e., going through the membership functions of the input data into fuzzy sets and their corresponding pertinence degrees to those sets, through a "fuzzy processing" or also called inference engine; and "defuzzification," i.e., it is a transformation from such a fuzzy evaluation into crispy variables, ready for control, or modeling analysis (Yen, 1999). Such a methodology could be compared to have a phasor domain analysis, or a Laplace domain analysis. Fuzzy data processing requires to be done in the fuzzy domain with proper direct and inverse transformations. Figure 4.1 shows the process of fuzzification, fuzzy inference engine, and defuzzification.

4.1 Fuzzification, defuzzification, and fuzzy inference engine

The mathematical structure to implement a transformation of real-world data into the fuzzy domain will need an inference engine (reasoning of input based on engineering domain or knowledge domain), a rule-base. Data must be converted to their real (crispy) values into fuzzy sets, evaluated by fuzzy rules, and the output could be either in a fuzzy statement, or in a real value. Such a domain transformation can be compared to using mathematical Laplace, or Z, or Phasor, engineering modeling analysis. Therefore, the fuzzy analysis for modeling and simulation has the following steps: (i) fuzzification, converting the physical variables into fuzzy sets and their respective membership functions; (ii) inference engine, which has a rule base with implication method to map fuzzy input sets into fuzzy output sets; and (iii) defuzzification, because most of the applications will need a real-valued crispy output.

4.1.1 Fuzzification

Fuzzification can be performed using many possible algebraic mathematical formulations for their membership functions, such as Gaussian, sinusoidal shaped,

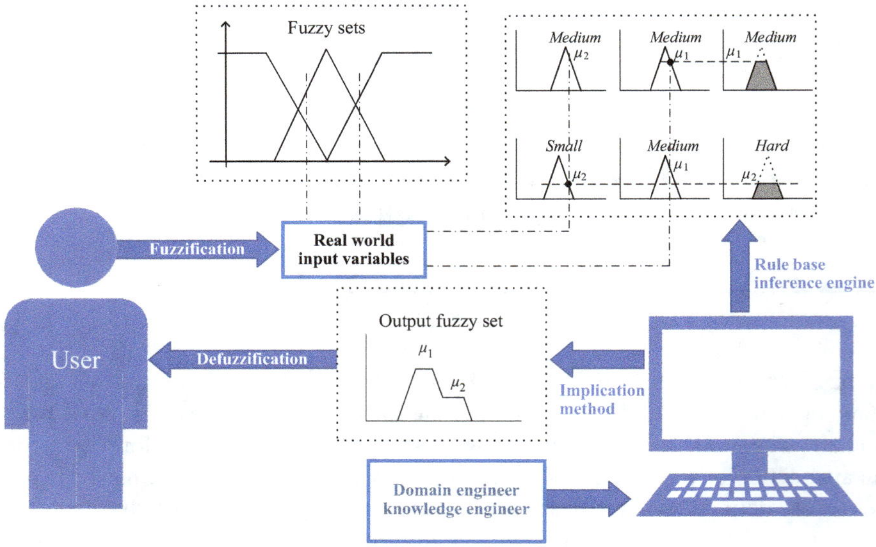

Figure 4.1 Fuzzy domain: there is a transformation "fuzzification" to have a decision made with a fuzzy-rule-based inference engine, then there is inverse transformation "defuzzification" to allow the user for support, or a process to be controlled, managed, or supervised

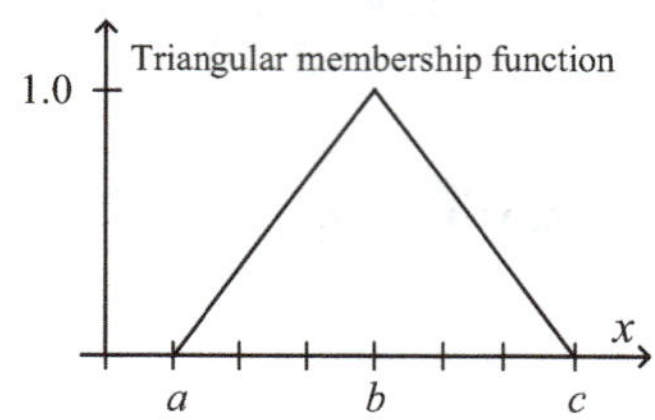

Figure 4.2 Triangular membership function, defined by rising and trailing linear segments and peak

sigmoid, continuous functions, and so on. Because fuzzification is the process of decomposing a system input and/or output into one or more fuzzy sets, and typically real-time processing and control are preferred in smart grid and power electronics applications, triangular or trapezoidal membership functions are the most common because they are easier to represent in embedded controllers. Figure 4.2 represents a typical triangular membership function, while Figure 4.3 represents a typical trapezoidal membership function. Each fuzzy set spans a region of input (or output) value graphed with the membership.

$$\mu(x; a \cdot b \cdot c) = MAX\left[MIN\left(\frac{x-a}{b-a}, \frac{c-x}{c-b}\right), 0\right] \tag{4.1}$$

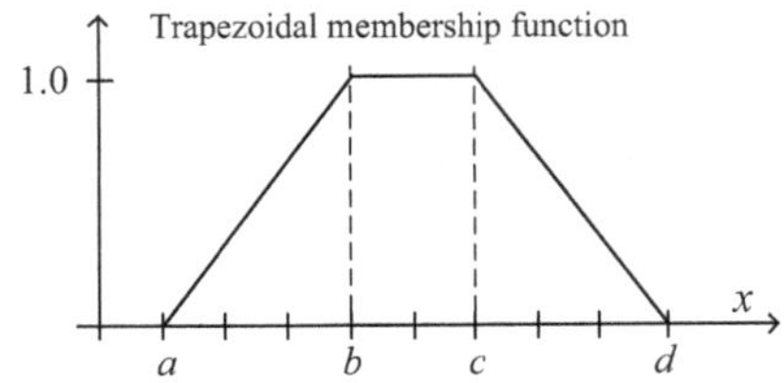

Figure 4.3 Trapezoidal membership function, defined by rising and trailing linear segments, and flat top plateau

$$\mu(x; a \cdot b \cdot c, d) = MAX\left[MIN\left(\frac{x-a}{b-a}, 1, \frac{d-x}{d-c}\right), 0\right] \tag{4.2}$$

Any particular input is interpreted from this fuzzy set, and a degree of membership is calculated. Equations (4.1) and (4.2) represent simple and direct implementation, just using straightforward calculations that can be implemented in any microcontroller or DSP hardware. The membership functions should overlap to allow smooth mapping of the system, which means any input variable will allow at least "firing" two fuzzy sets at the same time. The lack of overlapping can also be implemented in others to capture some nonlinearities, dead-band, or saturation of variables. The process of fuzzification allows the system inputs and outputs to be expressed in linguistic terms so that rules can be applied in a simple manner to express a complex system. Figure 4.4 shows an example for a variable with five or three fuzzy sets, all of which are triangular. Suppose a simplified implementation for an air-conditioning system with a temperature sensor. The temperature might be acquired by a sensor and microprocessor that has a fuzzy algorithm to process an output to continuously control the speed of a motor that keeps the room in a "good temperature." Such a microcontroller could in addition of controller the motor speed also direct a vent upward or downward as necessary for better air circulation.

Figure 4.5 illustrates the process of fuzzification of the input air temperature. There are five fuzzy sets for temperature: COLD, COOL, GOOD, WARM, and HOT. The membership functions for fuzzy sets COOL and WARM are trapezoidal, for GOOD is triangular, and for COLD and HOT are half-triangular with shoulders indicating the physical limits for such process (staying in a place with a room temperature lower than 8°C or above 32°C would be quite uncomfortable). The way to design such fuzzy sets is a matter of degree and depends solely on the designer's experience and intuition. Most probably Inuits, Yupiks, and Aleuts would disagree with an Equatorian choosing very different membership functions for such fuzzy sets. The figure shows some nonoverlapping fuzzy sets that can indicate any nonlinearity in the modeling process. There is an input temperature of 18°C that would be considered COOL with a degree of 0.75 and would be considered GOOD with a degree of 0.25. In order to build the rules that will control the air-conditioning motor, we could watch how a human expert would adjust the settings to speed up and slow down the motor in accordance to the temperature, obtaining the rules empirically. If the room temperature

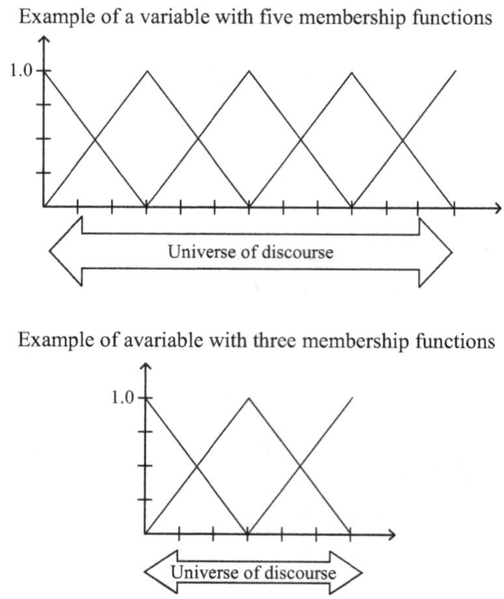

Figure 4.4 *Fuzzification example with only triangular membership functions, with either five or three fuzzy sets; the input fuzzification can easily be programmed in a user interface*

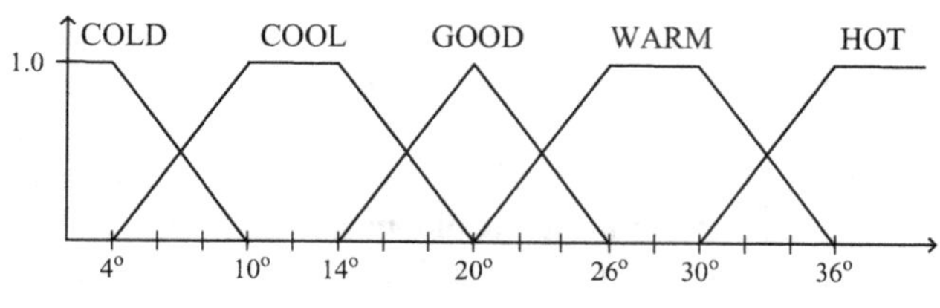

Figure 4.5 *Fuzzy sets defining temperature for air-conditioning management*

is good, keep the motor speed medium, if it is warm, turn the knob of the speed to fast, and blast the speed, if the room is hot. On the other hand, if the temperature is cool, slow down the speed, and stop the motor if it is cold. This is the beauty of fuzzy logic: to turn common sense, linguistic descriptions, into a computer-controlled system. Therefore, it is required to understand how to use logical operations to build the rules, and it is necessary to associate input fuzzy sets, through an inference engine, to generate output fuzzy sets. Such an inference is a mapping of an input range into an output range, in fact associating fuzzy sets in different and distinct universes of discourse.

4.1.2 Defuzzification

After the fuzzy reasoning, through an inference engine, we have a linguistic output variable. Defuzzification is the process of finding a crispy number that represents the information contained in such output fuzzy set, or the expected value of the solution. When the output of the fuzzy inference engine must be interpreted as a control action, or as a real value, for example, to configure a selector to go for an appropriated position, or to move a motor to a certain angle, or to rotate on a required speed, there are systems that do not need defuzzification because the fuzzy output would be interpreted in a qualitative way, for example, in manufacturing cheese, the fuzzy output would be compared to qualitative attributes as defined by humans who taste cheese, who use their taste, and smell to validate the quality of the cheese, and in such a situation a subjectively linguistic fuzzy semantic output would be reasonable.

Figure 4.6 shows five possible fuzzy sets to command the power for a motor controller, the universe of discourse is from −30 kW to +30 kW, with fuzzy membership functions such as ***Negative High***, ***Negative Medium***, ***Zero***, ***Positive Medium,*** and ***Positive High***. The figure also shows the output fuzzy set of a fuzzy inference engine. We can observe that "***zero***" and "***pos_med***" fuzzy sets are cut to a given height, making the total output fuzzy set as the overlapping combination of two trapezoids. Those heights are the strength of each rule associated with their particular output fuzzy set. A fuzzy inference engine has several rules in parallel, which is similar to using OR operation of those concatenated output rules. Each rule will have a rule strength factor, derived from the fuzzy operation on the antecedent (IF part). Suppose a fuzzy-logic-based crane controller defines the motor power for the crane, and Figure 4.6 shows that the rule strength giving "***zero***" is μ_1, while the other rule strength is μ_2 lopping off the "***pos_med***" fuzzy set. Other definitions for those inference engine rules' contribution are rule degree of validity, rule degree of strength, rule firing value, or rule degree of truth.

The output of the motor controller could assume the *center of area (CoA) (or gravity) of such an output fuzzy set,* indicated in Figure 4.6, by a hypothetical balancing of such a figure. The triangular block symbolizes a fulcrum that is trying

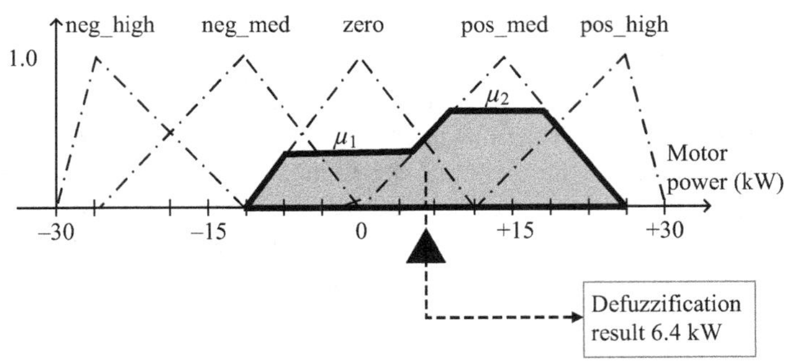

Figure 4.6 Fuzzy sets for a fuzzy-logic-based motor controller, indicating a combined overlapping trapezoidal shape of two fuzzy sets, with a fulcrum to quantify the defuzzification based on the center of gravity of such figure

to keep equilibrium of such figure, similarly to a traditional scale where one plate holds an object of given mass (or weight), and the scale achieves the equilibrium with the same mass on the other side. In Figure 4.6, the horizontal position of such a fulcrum will give the output, crispy, real valued, and command of 6.4 kW for the motor controller.

The objective of defuzzification is to derive a single real-valued numeric variable to best represent the fuzzy set inferred by an inference engine output. Therefore, defuzzification is an inverse transformation that maps the output from the fuzzy domain back into the crisp domain. The center of gravity (CoG) is also called the CoA method, or centroid, that computes the centroid of the composite area representing the output fuzzy term. There are many defuzzification techniques in the literature, but there are two prevailing methods: (i) composite moments, such as calculation of the CoG, CoA, or centroid, and (ii) composite maximum. The composite maximum is typically implemented as center of maximum (CoM), mean of maximum (MoM), or height method.

The CoM method is very popular with fuzzy controllers implemented on microcontrollers and RISC-based hardware. For the CoM, it is necessary to store only the peaks of the output membership functions, and the defuzzified crisp value is determined by finding the place where the weights are balanced. Therefore, areas and shape of output membership functions play no role, and only the maxima, i.e., singleton memberships are used. The CoM requires only the peaks of the membership functions, and the defuzzified value is determined by finding the fulcrum where the weights are balanced. This method is also called height method. Equations are very similar, except those for CoA where the areas of each membership function are used. For CoM, only their maxima are used. Naturally, the results are slightly different by using either the CoA (also called CoG) or the height method (also called CoM).

Equation (4.3) shows a composite moment implementation; such a method provides a crisp value based on the CoA of the fuzzy set. The total area of the membership function distribution used to represent the combined control action is divided into a number of subareas. The CoA method computes the centroid of the composite area representing the output fuzzy term (Figure 4.7). In the CoA defuzzification method, the fuzzy logic controller first calculates the area under the scaled membership functions and within the range of the output variable. The defuzzification module can use (4.3) in order to calculate the geometric CoA. CoA is the center of area, x is the value of the linguistic variable, and $x_{\min}$ and $x_{\max}$ represent the range of the linguistic variable. Equation (4.4) shows a discrete implementation, easier to implement in a microcontroller, RISC or DSP hardware. The area and the CoG or centroid of each subarea are calculated and then the summation of all these subareas is taken to find the defuzzified value for a discrete fuzzy set. Figure 4.8 shows a fuzzy set as an example in calculating their CoA (centroid) for defuzzification, with analysis displayed in (4.5).

$$CoA = \frac{\int_{x_{\min}}^{x_{\max}} f(x) \times x \, dx}{\int_{x_{\min}}^{x_{\max}} f(x) \, dx} \tag{4.3}$$

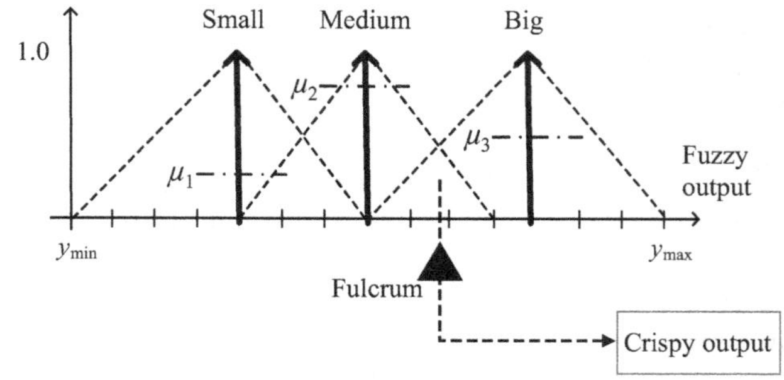

Figure 4.7 Output fuzzy sets defined as singletons: the rules will define each one their degree of strength, and a fulcrum will define the equilibrium point, allowing a defuzzified crispy output

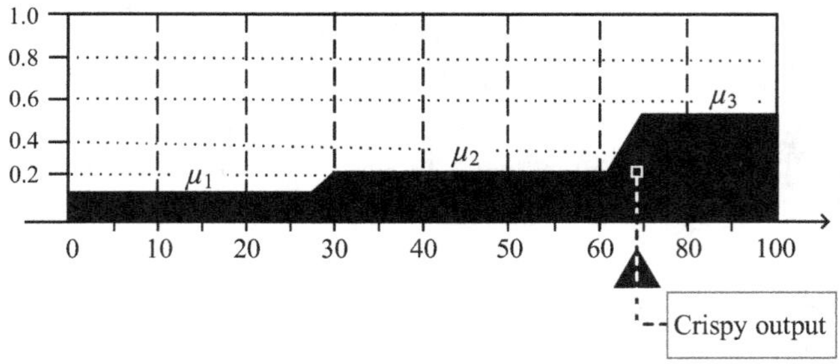

Figure 4.8 Example of a fuzzy set for the calculation of their centroid for defuzzification

$$u^* = \frac{\sum_{i=1}^{N} u_i \times \mu_{out}(u_i)}{\sum_{i=1}^{N} \mu_{out}(u_i)} \tag{4.4}$$

The defuzzification of fuzzy set in Figure 4.8 with CoA (centroid) is

$$\begin{aligned} CoA &= \\ &= \frac{(0+10+20)\times 0.1 + (30+40+50+60)\times 0.2 + (70+80+90+100)\times 0.5}{0.1+0.1+0.1+0.2+0.2+0.2+0.2+0.5+0.5+0.5+0.5} \\ &= 67.4 \end{aligned} \tag{4.5}$$

The **modified CoA** is implemented when a fuzzy logic controller considers the full area under the scaled membership functions, even if this area extends beyond the range of the output variable; in such modified method, interval of integration is from the minimum membership function value to the maximum membership function value, considering that such interval may extend beyond the range of the output variable. Another possibility for modification is called **center of sums**, in which the fuzzy logic controller first calculates the geometric CoA for each membership function; then it calculates the weighted average of the geometric CoA for all membership functions, where there is each and every geometric center of each membership function of the output fuzzy set; they are scaled membership function and each area is scaled by their degree of strength coming from the fuzzy inference engine implication.

The **CoM** requires only the peaks of the membership functions. The defuzzified value is determined by finding the fulcrum where the weights are balanced. This method is also called height method. The crisp output is computed as a weighted mean of the term "membership maxima," weighted by the inference results. Equation (4.6) is very similar to the centroid one (4.4), except the CoA where the areas for all membership functions are used, while for CoM only their maxima are used. The typical numerical value is the mean of the numerical values corresponding to the degree of membership at which the membership function was scaled. The CoM and CoA defuzzification methods usually apply to closed-loop control applications of fuzzy logic. These methods typically result in continuous output signals because a small change in input values does not change the best compromised value for the output.

$$u^* = \frac{\sum_{i=1}^{N} u_i \times \sum_{k=1}^{n} \mu_{O,k}(u_i)}{\sum_{i=1}^{N} \sum_{k=1}^{n} \mu_{O,k}(u_i)} \tag{4.6}$$

In the **MoM** defuzzification method, the fuzzy logic controller first identifies the scaled membership function with the greatest degree of membership. The fuzzy logic controller then determines the typical numerical value for that membership function. The typical numerical value is the mean of the numerical values corresponding to the degree of membership at which the membership function was scaled. The MoM defuzzification method is usually employed in pattern recognition applications. It approaches the most plausible result, rather than averaging the degrees of membership of the output linguistic terms; the MoM defuzzification method selects the typical value of the most valid output linguistic term.

4.1.3 Fuzzy inference engine (implication)

The fuzzy inference engine concatenates the model of the system to be controlled or modeled; it consists of a database (linguistic membership functions) plus a rule base with linguistic rules. The database allows definitions used in a set of fuzzy rules, while the rule base categorizes the objectives and the strategy of control used by people with expertise, or it allows a model construction based on technical expertise and understanding of the system, or it may allow a hybrid construction partially fuzzy and partially based on multi-parametric linear equations. When the rule base has fuzzy rules based on the logic of fuzzy inputs with outputs based on fuzzy sets, such inference engine is called "***Mamdani Fuzzy Inference***," or "***Type 1 Fuzzy Inference***." On the other hand, when the rule base has fuzzy rules based on logic of fuzzy inputs, with each rule giving an output of a linear equation based on the input data, that is a hybrid fuzzy modeling, called "***Takagi–Sugeno Fuzzy Inference***," or "***Parametric Fuzzy Inference***," or even "***Type 2 Fuzzy Inference***," this inference engine is very powerful because it associates a fuzzy understanding of input variables with approximated linear modeling, such as dividing a complex nonlinear problem in linearized sections and combining it to form an output, when enhanced with recursive-least-squares, it becomes a multi-parametric recursive fuzzy modeling applicable for nonlinear, dynamic, time-varying problems, capable of real-time adaptive performance.

Which defuzzification method should be chosen could be based on understanding the nature of the modeling or control approach. Techniques for inference engine interaction with defuzzification schemes suggest that when the implication method is correlation minimum and min/max inference, the best choice for defuzzification is a composite maximum; on the other hand, when the implication method is correlation product and additive inference, the best choice for defuzzification is composite moments. Other possible implications and techniques can be used, if additive inference tends to smooth out the plateaus caused by the correlation minimum techniques.

In closed-loop control applications, it is very important to have smooth output with continuous functions, because if the output of a fuzzy controller has sharp variations, or discontinuities, that may cause instability and oscillations in the overall closed-loop behavior. Therefore, for closed-loop control is probably best to adopt CoM defuzzification. When a fuzzy PI or fuzzy PID is implemented, there is an integrator in the output of the controller, and the process will still receive a continuous function, even using the MoM defuzzification. If the controller has no embedded integration, the choice of defuzzification must be done only on the basis of smooth output signals. For pattern recognition, it can be used for the MoM defuzzification, because for the classification of clusters it is better to use the most plausible pattern; the possibility vector is the result of the classification, containing information on similarity of the inferred output with objectified patterns, similar to a probability density function. When a fuzzy system is used for supporting decision-making, the choice of the defuzzification method depends on the context of the decision. For quantitative decisions, such as resource allocation, project prioritization, levels of resources for manufacturing or process control, it is recommended to use CoM. If the supporting fuzzy decision-making is used for qualitative decisions, such as fraud detection in card transactions, credit evaluation, insurance evaluation, electric power distribution safety, and energy management transactions, it is recommended to use MoM. The fuzzy inference engine and types of implication are discussed in Section 4.2.

4.2 Fuzzy operations in different universes of discourse

Statements assert facts or states of affairs. They give descriptions that can be organized in several rules of thought. Application of mathematical descriptions and the use of logical rules for formulating hypotheses were developed by several philosophers. They have been influenced by the earlier syllogistic logic, in which premises were manipulated to produce true conclusions. A typical syllogistic rule of inference that goes by the Latin name modus ponens (affirmative mode) can be given as

If A is true
AND A implies B
THEN B is true

where the connectives AND, OR, and NOT are essential operations to derive the truth of such reasoning. Such logical connectives are defined for fuzzy logic operations, and they are closely related to Zadeh's definitions of fuzzy sets operations. However, when the relations of fuzzy sets A and B are on a different universe of discourse, the implication R denotes the relation between the fuzzy sets A and B with an implication.

$$\textbf{Implication } R{:}(x = A) \rightarrow (y = B) \; \textbf{\textit{IF}} \; x \text{ is } A \; \textbf{\textit{THEN}} \; y \text{ is } B \tag{4.7}$$

A fuzzy implication is an important connective in fuzzy control systems because control strategies can be written with IF-THEN rules. There are various types of techniques in which fuzzy implication may be defined. Such relationships are mostly derived from multivalued logic theory. The followings are some of the common techniques of fuzzy implication that can be used in practice. Note that Mamdani's implication is equivalent to Zadeh's implication when $\mu_A(x) \geq 0.5$ and $\mu_B(y) \geq 0.5$.

Zadeh's implication:

$$\boldsymbol{\mu_R(x,y) = max\{min[\mu_A(x), \mu_B(y)],\ 1 - \mu_A(x)\}} \tag{4.8}$$

Mamdani's implication:

$$\boldsymbol{\mu_R(x,y) = min[\mu_A(x), \mu_B(y)]} \tag{4.9}$$

A fuzzy set $A \subset E_1$ will induce another fuzzy set $B \subset E_2$ the membership function of which will be $\mu_B(x/y)$. When we have a multi-input–multi-output system with many rules, characterized by a chain of statements such as

IF var1 = A ⟨**connective**⟩ var2 = B ⟨**connective**⟩ … **THEN** Out1 = C ⟨**connective**⟩ … ⟨**connective′**⟩

IF var1 = D ⟨**connective**⟩ var2 = E ⟨**connective**⟩ … **THEN** Out2 = E ⟨**connective**⟩ … ⟨**connective′**⟩

… …

… …

where A, B, C, D, and E are either crisp or fuzzy sets, and ⟨***connective***⟩ represents the particular fuzzy operator chosen to express the fuzzy inference or fuzzy implication desired. We can define the rule base as a relation $\mathcal{R}$ and the output fuzzy set could be given by the compositional operator " ∘ ." A compositional rule of inference, which for the purpose of practical computation, can be written in terms of the membership functions of their respective fuzzy sets, is indicated in (4.10), with max–min indicated in (4.11), and max-product indicated in (4.12).

$$B(y) = A(x) \circ \mathcal{R}(x, y) \tag{4.10}$$

Max–min composition

$$\mu_B(y) = \mathbf{max}[\mathbf{min}(\mu_A(x), \mu_R(x, y))]\ x \subset E_1 \tag{4.11}$$

Max-product composition

$$\mu_B(y) = \mathbf{max}[\mathbf{min}(\mu_A(x) \times \mu_R(x, y))]\ x \subset E_1 \tag{4.12}$$

4.3 Mamdani's rule-based Type 1 fuzzy inference

In a fuzzy modeling or a fuzzy control system, the expertise is embedded in a set of rules of inference, such as **IF** ⟨**conditions**⟩ **THEN** ⟨**consequences**⟩. Just as an example, suppose a fuzzy system has two rules, such as

Rule 1 IF the distance between two cars is *medium* AND the speed of the car is *medium* THEN brake is *medium* for speed reduction.

Rule 2 IF the distance between two cars is *small* AND the speed of the car is *medium* THEN brake is *hard* for speed reduction.

The terms "medium," "small," and "hard" are linguistic variables that will be defined in their fuzzy sets with their corresponding membership functions. Figure 4.6 shows how two rules can be combined into a decision-making strategy, i.e., a fuzzy output set, which can be converted next into a real-valued number to command the braking action of a car. Rule-based fuzzy controllers have several advantages; the fuzzy control rules are easy to understand by maintenance personnel and control functions associated with a rule that can be tested individually. This improves maintainability because the simplicity of the rules allows the use of less skilled personnel. Individual rules can combine to form a structured complex control. Parallel processing allows fuzzy logic to control complex systems using simple expressions. In addition, rules can be added for alarm conditions, both linear and nonlinear control functions can be implemented by a rule-based system, using expert knowledge formulated in linguistic terms. Eventually, fuzzy controllers are inherently reliable and robust because a partial system failure may not significantly degrade the controller's performance.

Several rules in parallel are the same as using an OR operation of those concatenated output rules. Each rule will have a rule strength factor, derived from the fuzzy operation on the antecedent (IF part). For example, in Figure 4.9, the first rule strength is μ_1, while the second rule strength is μ_2; since the rules are in parallel, the

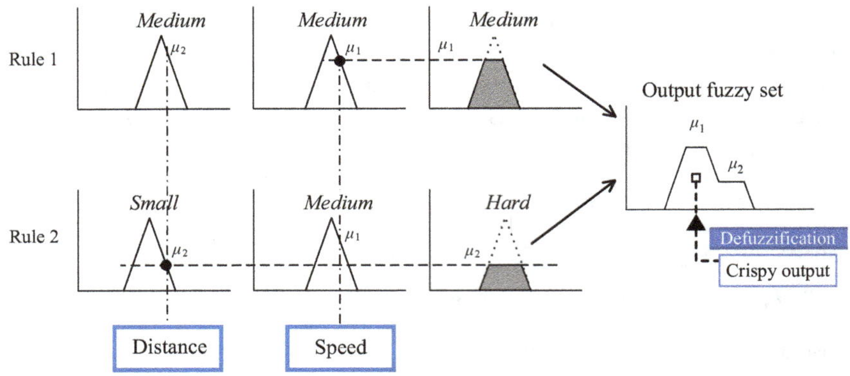

Figure 4.9 Inference of fuzzy rules for controlling the brake of a car when considering speed of the car and distance to the leading one. The max–min composition gives a fuzzy set at the output, which can be converted to a crispy variable using the center of gravity of that geometrical figure

OR of those gives the fuzzy set as indicated on the right side of the figure. Of course, there is some overlap of one fuzzy term from one rule with another one: depending on the defuzzification method, this overlap might be considered or not. The output of this system could be, for example, the CoG of the fuzzy set output.

4.4 Takagi–Sugeno–Kang (TSK), Type 2 fuzzy inference, parametric fuzzy, and relational-based

Takagi and Sugeno introduced an inference structure based on fuzzy sets theory (Takagi and Sugeno, 1985). Such structure has several names; it is common to call such an inference by Takagi–Sugeno, or TS fuzzy inference, or maybe it is called a parametric or relational fuzzy system, or Type 2/Type II fuzzy system. Eventually, after original Sugeno's proposal there was further development by Kang (Takagi and Sugeno, 1985; Sugeno and Kang, 1988), so it has been called in the past few years as Takagi–Sugeno–Kang (TSK) fuzzy inference.

In contrast to what has been previously discussed, i.e., the fuzzy rule base structure, or Mamdani-type or Type 1 fuzzy system, we can simplify as just saying Type 1 or Type 2 fuzzy control. Type 2 fuzzy systems are based on a rule base approach only to evaluate the antecedents, after defining the rule strength, each consequent will be a linear parametric equation (instead of fuzzy sets), in terms of the inputs of the system. In a TSK fuzzy modeling or a fuzzy control system, the expertise is embedded both a set of rules of inference plus linear equations such as **IF ⟨conditions⟩ THEN ⟨linear equation of inputs⟩**. Just as an example, the rule of (4.13) defines that when x_1 is small and x_2 is big, the value of y would be to the sum of x_1, x_2 plus $2x_3$, where x_3 is an input variable for the system but not conditioned in the premise. Equation (4.14) illustrates a general TSK rule.

$$\mathbf{R{:}IF}\ \boldsymbol{x_1}\text{is } \textit{SMALL}\ \mathbf{AND}\ \boldsymbol{x_2} \text{ is } \textit{BIG}\ \mathbf{THEN}\ \boldsymbol{y} = \boldsymbol{x_1} + \boldsymbol{x_2} + \mathbf{2}\boldsymbol{x_3} \tag{4.13}$$

$$\textit{TSK Rule}: \mathbf{IF}\ \boldsymbol{x_1} \text{ is } \boldsymbol{A_1}\ \mathbf{AND}\ \boldsymbol{x_2} \text{ is } \boldsymbol{A_2} \cdots\ \mathbf{AND}\ \boldsymbol{x_k} \text{ is } \boldsymbol{A_k}\ \mathbf{THEN}\ \boldsymbol{y} = \boldsymbol{b_0} + \boldsymbol{b_1 x_1} + \cdots + \boldsymbol{b_k x_k} \tag{4.14}$$

We should only use AND connectives in the premise and adopt a linear function in the consequence. For each input x_k there is a multiplicative coefficient parameter b_k, and this is the reason why TSK fuzzy inference is also called parametric fuzzy inference. Figure 4.10 shows an example of two TSK fuzzy rules with two linear equations, generating an output variable based on the weighted average method of defuzzification; (4.15) shows the general implementation of an output coming from the concatenation of all fired TSK fuzzy rules.

$$y = \frac{\sum_{r=1}^{n} \mu_r \times f_r(x_1^*, \quad \ldots \quad , x_m^*)}{\sum_{r=1}^{n} \mu_r} \tag{4.15}$$

The TSK, or parametric fuzzy inference engine, provides a powerful tool for synthesizing a model of highly nonlinear functional mapping, as depicted in Figure 4.11, where three fuzzy sets for one input variable will provide three rules. Each rule will divide the system space in several trajectories with several multi-linear parametric functions. The design issues will be to define the AND connectives plus the parameters, which are usually done with input/output data and multi-regression linear analysis. In fact, the consequences could be any function,

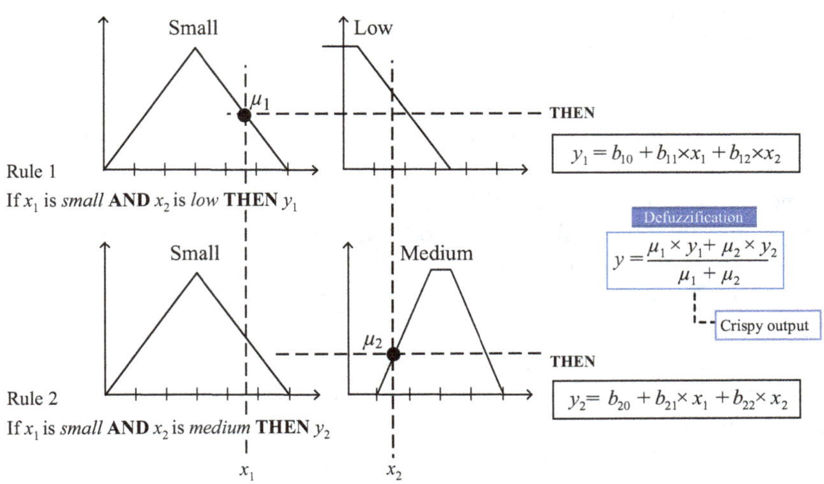

Figure 4.10 Two parametric fuzzy rules (TS) relating variables x_1 and x_2 are with linguistic variables (small, low, and medium) with output equations for each region, the composed output of two rules is calculated with the weighted average method of defuzzification

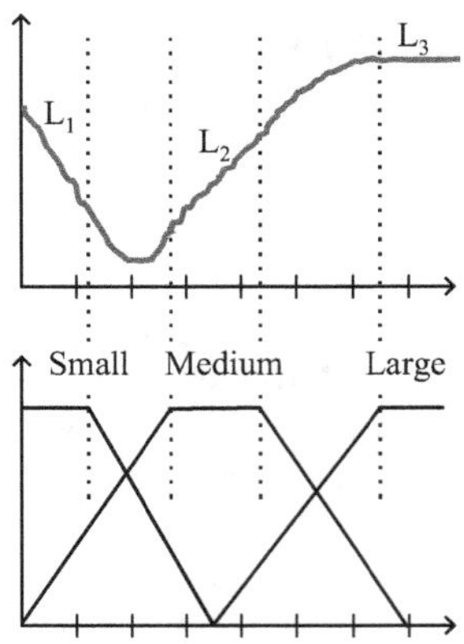

Figure 4.11 Example of a TSK fuzzy model, expressing a highly nonlinear functional relation with a few fuzzy sets and interpolation of linearized sections

polynomials, or even nonlinear. However, linear functions have been employed most of the time. Off-line linear regression can be easily implemented in MATLAB®, or in an Excel spreadsheet. Adaptation can be implemented with adjusting training associated with recursive-least-squares methodology. The main feature of a TSK model is as it is hybrid, fuzzy evaluation of input data could be associated with heuristics, helping one to merge the nonlinear description with real-time adaptive control. Typically, a TSK inference engine will require less rules than a Mamdani's inference engine. The combination of a global rule-base description with local linear approximations by means of a linear regression model corresponding to a linear input–output model would make such a TSK fuzzy inference for describing the system locally. Of course, it is necessary to gather and arrange input/output data for performing a multivariable linear regression for each rule.

One example for using a TSK method has been discussed in the study of Simoes and Kim (2006) for fuzzy modeling approaches for the prediction of machine utilization in hard-rock tunnel boring machines (TBMs), application in Civil and Geological Engineering. A very complex machine with expensive operation and maintenance is used for boring tunnels. The operating cost and maintenance is extremely high, and any hour savings might be important for the cost of the project. Adjustments on the advance of the machine depends on the boring machine diameter (MD), the kind of rock (rock mass rating (RMR)), if there is inflow of ground water, and what is the rock quality designation (RQD). In the study of Simoes and Kim (2006), a model was designed with four inputs; MD has three membership functions, while RMR has four, groundwater inflow (GI) has four, and RQD has three. If a Type 1 fuzzy rule base was used, it would be necessary that the evaluation of $3\times4\times4\times3=144$ rules. It was found that with less equations, the designer would define the patches of the input/output mapping,

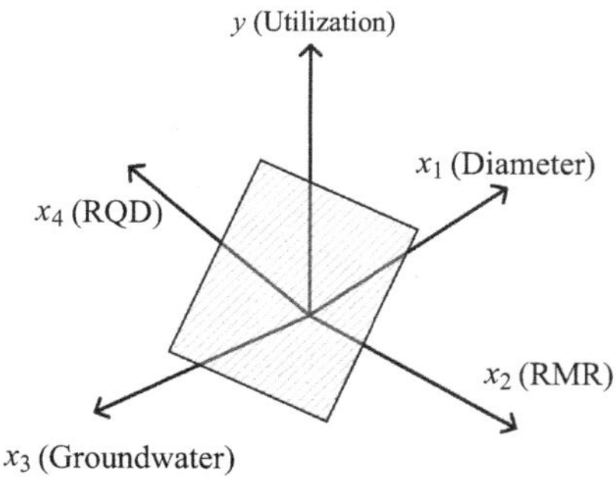

Figure 4.12 Four input variables can be considered in a tunnel boring machine fuzzy management to define a linear hyperplane of the utilization and advance of the machine

resulting in a smaller Type 2 (TSK) fuzzy rule base. There are several ways to perform a multivariable linear regression in order to find the equation linear coefficients (or parameters). Suppose that a function is supposed to be fitted as a linear equation defined by the multiplicative parameters (linear coefficients). That can be visualized as a linear function of a hyperplane, as illustrated in Figure 4.12, where it is possible to the "least square method" in order to calculate the best fitting straight line, using the equations indicated in (4.16) with (4.17)–(4.21).

$$y = b_1x_1 + b_2x_2 + b_3x_3 + b_4x_4 + b_0 \tag{4.16}$$

$$m = \frac{n(\sum xy) - (\sum x)\cdot(\sum y)}{n(\sum (x^2)) - (\sum x)^2} \tag{4.17}$$

$$b = \frac{(\sum x)(\sum (x^2)) - (\sum x)\cdot(\sum xy)}{n(\sum (x^2)) - (\sum x)^2} \tag{4.18}$$

$$\sum_{i=1}^{n} (y_i - \bar{y}_i)^2 = \sum_{i=1}^{n} (y_i - \hat{y}_i)^2 + \sum_{i=1}^{n} (\hat{y}_i - \bar{y}_i)^2 \tag{4.19}$$

$$SST_{(total\ sum\ of\ square)} = SSE_{(sum\ of\ square\ for\ error)} + SSR_{(sum\ of\ square\ for\ regression)} \tag{4.20}$$

$$R^2 = \frac{SSR}{SST} = 1 - \frac{SSE}{SST} \tag{4.21}$$

A multivariable linear regression can be made off-line in MATLAB or with an Excel spreadsheet, as long as enough input/output data are provided for each group of linear equations, a lot of data are necessary for such training. The ***LINEST*** Excel function calculates the statistics for a straight line that explains the relationship between the independent variable and one or more dependent variables and returns an array describing the line.

The rule-based Mamdani inference engine has been compared with the TSK-parametric-based (Sugeno model) fuzzy engine to model TBM datasets in the study of Simoes and Kim (2006). A total of three hard TBM projects were studied to establish possible trends and correlations between rock mass properties and machine utilization. Since rock mass properties are the most affecting and unpredictable factors to machine utilization, only rock mass properties were focused and analyzed in this chapter. The identification of input parameters includes MD, RMR, GI rate, and RQD. These were used as input parameters influencing machine utilization level for both algorithms. In order to verify the validity of the two models, the predicted machine utilization level and the measured (or real) utilization level from the field records were compared. The TSK model was a more accurate estimator of machine utilization than Mamdani's, with a smoother resolution. By applying this utilization predictor model for the planning stage of TBM projects, a machine advance rate and corresponding total excavation time and cost were then estimated and used for TBM project planning, management, and bidding purposes.

4.5 Fuzzy model identification and supervision control

A rule-based fuzzy approach (Type 1) is more suitable for acquiring and implementing expert human operator knowledge, while the parametric fuzzy approach (Type 2) is best used when input/output numerical data are available; in addition, the parametric fuzzy approach yields better estimation accuracy because it is a hybrid of rule-based fuzzy and numerical components. However, the rule-based fuzzy approach requires no training, while the parametric fuzzy approach requires linear coefficient adjustment performed by statistical multi-linear procedures. Fuzzy control has lot of advantages when used for optimization of alternative and renewable energy systems. The parametric fuzzy algorithm is inherently adaptive, because the coefficients can be altered for system tuning. Thus, a real-time adaptive implementation of the parametric approach is feasible by dynamically changing the linear coefficients by means of a recursive-least-square algorithm repeatedly on a recurrent basis. Adaptive versions of the rule-based approach changing the rule weights (degree of support) or the membership functions recurrently are possible. The disadvantage of the parametric fuzzy approach is the loss of the linguistic formulation of output consequents, sometimes important for industrial plant-process control environment.

There is another important inference engine, called SAM—standard additive model, that is a generalized inference model proposed by Kosko (Fuzzy Engineering). The additive structure comes from the summation of fired THEN sets, which is based on the sup-product composition and the use of addition as a rule aggregation operator (Yen, 1999). The SAM consists in a fuzzy model composed of N parallel rules, antecedents and consequents of which are fuzzy sets, although it uses fuzzy sets in the inference engine, SAM is similar to that used in the TSK model, considering that the linear equation has one coefficient that is a fuzzy set where a fuzzy conclusion of the model output will consider a scaling approach instead of a clipping method. The output set has its membership function cutoff in the top, α—cut value of which is equal to the degree of firing for that rule. On the other hand, in the scaling method the membership function is scaled down in the proportion of the degree of firing (Yen, 1999). In the SAM model, the inputs are necessarily crisp numbers and the inference procedure produces an output fuzzy set that must be defuzzified by the centroid (CoA) method.

Most controllers in operation today have been developed using conventional control methods. There are, however, many situations where these controllers are not properly tuned, and there is heuristic knowledge available on how to tune them while they are in operation. There is then the opportunity to utilize fuzzy control methods as the supervisor that tunes or coordinates the application of conventional controllers. Industry uses most of 90% of PID controllers, because they are easy to understand, easy to explain to others, and easy to implement. Moreover, they are often available at little extra cost since they are often incorporated into the programmable logic controllers that are used to control many industrial processes. Unfortunately, many of the PID loops that are in operation are in continual need of monitoring and adjustment since they can easily become improperly tuned; due to plant parameter, or variations in operating conditions, there is a significant need to develop automatic tuning of PID controllers, particularly keeping the process or plant still in operation. A fuzzy supervisor management system can be made adjustable for the PID gains, providing the human operator with an indication that there will be different effects on the control system that will cause it to become out of tune. A "behavior recognizer" seeks to characterize the current behavior of the plant in a way that will be useful to the PID designer. The whole supervisor may be implemented as an adaptive controller with the following tuning rules:

- if steady-state error is large then increase the proportional gain,
- if the response is oscillatory then increase the derivative gain,
- if the response is sluggish then increase the proportional gain,
- if the steady-state error is too big then adjust the integral gain, and
- if the overshoot is too big then decrease the proportional gain.

Fuzzy and neuro-fuzzy techniques became efficient tools in modeling and control applications; there are several benefits in optimizing cost effectiveness because fuzzy logic is a methodology for the handling inexact, imprecise, qualitative, and fuzzy verbal information in a systematic and rigorous way. A neuro-fuzzy controller generates, or tunes, the rules or membership functions of a fuzzy controller with an artificial neural network approach. For applications of alternative and renewable energy systems, it is very important to use artificial intelligence techniques such as fuzzy logic and neural networks, because the installation costs are high, the availability of the alternative power is by its nature intermittent, and the system must be supplemented by additional sources to supply the demand curve. There are efficiency constraints, and it becomes important to optimize the efficiency of electric power transfer, even for the sake of relatively small incremental gains, in order to amortize installation costs within the shortest possible time. Several ANFIS (artificial neural network fuzzy inference systems) use the TSK approach, and the coefficients are trained with backpropagation or gradient descent, instead of multivariable linear regression.

Chapter 5
Fuzzy-logic-based control

Fuzzy modeling and control approaches can be categorized as: (i) fuzzy reasoning (or knowledge) systems or fuzzy decision-making systems, and (ii) fuzzy modeling/control systems. Those categories use fuzzy logic with specific requirements. Fuzzy reasoning systems may arrive a qualitative knowledge for a given problem, for example, a fuzzy expert system may help a health provider to define some emergency treatment in cases of trauma or immunological breakdown, with procedures and expertise based on natural language. In such a case, there is no need for defuzzification because qualitative analysis can be implemented with fuzzy sets in mapping qualitative facts. Linguistic results would convey enough information for such an intelligent system. A different strategy must be used for fuzzy modeling or fuzzy closed-loop control. Fuzzy modeling requires variables from real world as input, fuzzified, modeled under fuzzy rules, and the output of this model might be most often in real-valued numbers, for example, a nonlinear compressor for fuel cells might have ill-conditioned nonlinear differential equations, but either using data (for Takagi–Sugeno–Kang (TSK) method), or an expert description (for Mamdani's method), will allow a model to associate temperature, pressure, airflow, and hydrogen flow into fuel cell electrical energy output and thermal energy. Fuzzy controllers will always need a crisp value as result, because such a control action would have to be translated to a physical actuator, for example, if a certain valve has to be "opened somewhat" that is not a useful output, and defuzzification is required. Fuzzy sets are used as a convenient tool to define control rules and to make inferences. But at the end, a closed-loop control must take crisp inputs, process through a fuzzy inference engine, and should eventually compute a crisp output.

> "As complexity rises, precise statements lose meaning and meaningful statements lose precision"—Lotfi A. Zadeh

When a microprocessor, or microcontroller, or a DSP is used in computer control applications, sample-and-hold circuits are inserted at the digital-to-analog interfaces. The simplest device available is a zero-order hold that holds the output constant at the value fed to it at the last sampling instant (a piecewise constant signal is generated). Higher order holds are also available, which use a number of previous sampling instant values to generate the signal over the current sampling interval. In a digital control loop, the following procedure must take place: (i) measure system output and compare with the desired value to give an error; (ii) use the error, via a control law, to compute an actuating signal; (iii) apply this corrective input to the system; (iv) wait for the next sampling instant; and (v) repeat this algorithm in a constant high frequency greater than the system's bandwidth response.

5.1 Fuzzy control preliminaries

Figure 5.1 shows the conventional methodology in designing control systems. What is modeled is the plant or process that will be controlled. This procedure is called system identification, where the plant or process is assumed to be linear, or approximately linear, characterized with differential equations; if those can be solved, an analytical transfer function can be derived, and then parameters of a controller, or regulator, in a closed loop will have to be adjusted for an adequate behavior, such as maximum overshoot, damping, velocity response, accommodation time, and steady-state error.

There are several systems that are not best implemented with traditional automatic control, particularly when human operators are in the control loop and the methodology is focused on how those operators observe the system, adjust set points, and control the quality in a heuristic way. In a bakery, or in a sauce factory, or maybe in a beer production plant, the behavior of the operator is what matters, how they will adjust parameters, temperatures, mixing strength, and humidity for a given set of circumstances.

Several relevant contributions by different researchers implemented energy management systems in microgrids, proposing different control strategies (Sreelekshmi *et al.*, 2017; Azizi *et al.*, 2019; Oliveira *et al.*, 2015; Liu *et al.*, 2019; Khavari *et al.*, 2020; Pascual *et al.*, 2014; Zaheeruddin and Manas, 2015; Iovine *et al.*, 2017; Nnaji *et al.*, 2019; Abrishambaf *et al.*, 2017). It is also possible to find authors who applied fuzzy logic to manage the demand (Sharma *et al.*, 2018; Badwawi *et al.*, 2019; Lebrón and Andrade, 2016; Oliveira *et al.*, 2017; Bhowmik *et al.*, 2018; Angalaeswari *et al.*, 2017; Chekired *et al.*, 2017). However, the number of researchers using fuzzy logic to control energy management system in microgrid and presenting financial cost savings is relatively low (Khalid *et al.*, 2019; Banaei and Rezaee, 2018; Simões *et al.*, 1997b; Fossati *et al.*, 2015). Mansiri *et al.* (2018) proposed a fuzzy-logic-based algorithm utilizing the time of using

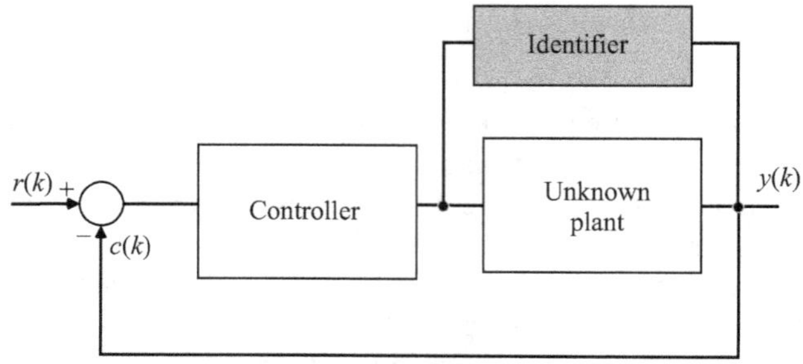

Figure 5.1 Identification of process or plant dynamics

electricity-pricing concept to minimize the cost operation of a smart grid. The system proposed can strategically handle fluctuating PV production alternating between absorbing power during high solar irradiation periods and discharging energy to the load during peak consumption times. The results show that the algorithm successfully decreased the annual electricity bill.

In this chapter, we have a simple fuzzy logic control methodology that is applicable for most of systems. It has been proven and utilized for the first time in renewable energy systems and wind turbine control systems in 1997 (Simões *et al.*, 1997a,b; Souza *et al.*, 1997), when it was developed in SIMNON, a DOS-based simulation environment for nonlinear systems, the code was written in C and compiled to run on a TI C30 DSP hardware implemented with a double back-to-back pulse-width modulation (PWM) control for an induction generator.

Figure 5.2 shows the fuzzy methodology where the operator is the model to be identified, while she or he is controlling the system. Therefore, a fuzzy controller has its fuzzy model mimicking the human expert who controls, handles, and manages the system. This paradigm shifts from the point of view of "modeling the physical system" into "modeling the human operator," and it changes dramatically how automatic control can be implemented in modern computers. The use of an intelligent control, such as fuzzy logic or neural network based, cannot be seen as a universal panacea, and it is not the correct attitude of simply changing a conventional controller by a fuzzy controller, just believing it would present the best performance. There are yet analyses to be done after deciding on such fuzzy-logic-controller-based design. If the plant or process to be controlled is not completely linear, but the nonlinearity is known and it is a smooth function, no discontinuities, a regular or classic-based control design is still the best option. Another possibility is when the process or system is very highly nonlinear, but it may operate in a quiescent operating point, where only small variations happen, and certainly a Taylor function expansion would allow a small-signal averaged model, classic and modern control theory can also be better applied.

Typically, solutions with a proportional–integral–derivative (PID) control are the most reasonable to design, or maybe state-variable feedback control (with

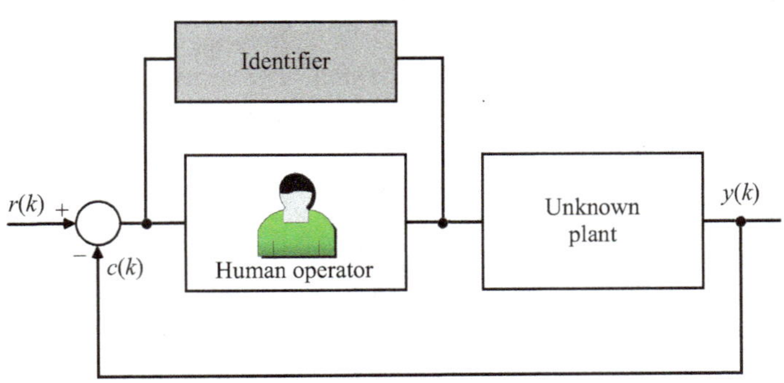

Figure 5.2 Identification of control operator behavior

state-space theory), or analog systems with lead, lag, and lead–lag regulators, could be the solution to be adopted. PID controllers can usually control process and plants, even with unknown dynamics, because the P component represents the instantaneous feedback error, I component represents the integral and serves as a memory of the retro-feedback, and the D component that is a derivative of error anticipates the future of the feedback control. Someone with expertise can look at those terms and manually fine-tune the PID parameters or use some classic control-based design that will have a very damped response, maybe not optimal, but still stable for highly nonlinear plants. If the controller parameters are tuned, the performance of the closed-loop control will be satisfactory, maybe not optimal, but still arriving to zero-steady-state error if possible. The task of tuning requires the multiple observation and achieves the most perfect, functional, or effective as possible, by considering damping, overshoot, settling time, offset, steady-state error, reaction to parameter variation, and reaction to step transitions in the set points. A designer will assume that three individual strategies (P+I+D) are decoupled and can be added, making the closed-loop control to compensate for parameter variation, noise, and environmental alterations, but to some complex nonlinear plants, those real-life affects are sometimes not possible to be considered as combining in a linear model perspective. A nuclear power plant may be considered as one set point or reference such as "amount of power to be generated" with an output such as "electrical power generated ready for transmission"; however, such a process is so complex, so complicated, with so many inner loops, so many internal subsystems, so many possibilities of faults and reliability issues, that a simple PID control will never be able to be applied for a single-input/single-output simplified version of such a system. Heavy mathematics and signal processing may help, but then the understanding of control becomes blurred. The history shows that human beings were able to make nuclear power plants to operate even without using complicate computer models or heavily theoretical approaches, by having layers of control, where experts help with their understanding, starting from the fission of the nuclear fuel, to steam and turbines expansion, into having all the signals to indicate safety and integrity of a controlled nuclear reaction to generate electrical power.

Supervisory control systems are very important, because they allow decoupling of complex systems into feasible smaller tasks. A brewing company, or an alcohol production for sugarcane, or a cement and concrete factory production, or sewer treatment, or a wind farm, a large PV array, a fuel cell, all these systems can be controlled and managed with supervisory control systems that allow experts to look at variables, instrumentation, closed-loop control error responses, change to set points, and make real-time fine tuning. Their expertise can be captured with fuzzy control rules, by modeling the operators, instead of the process. Fuzzy parametric relational (TSK) can be implemented by fuzzy evaluation of input data and multi-parametric linear expansion of equations to be implemented in fuzzy control rules.

In several industrial processes, such as extrusion, rubber, elastomers, tires, Banbury mixer, fermentation, distillation, ceramics, ferrites, permanent magnets, and food mills, there are no mathematical functions describing their input/output transfer functions, because of complex reactions, dynamics not formalized, interrelated multidisciplinary complex models, and unknown parameters, and expert people still work with these processes. There are empirical recipes, accepted procedures, and adopted parameters to produce acceptable results. Such recipes could be taken as input/output functions, because they relate input variables that are fundamental for those processes to obtain the desired outputs, even if dimensionless variables are used, such as rubber composition and origin of the rubber tree with a thickness of tire and lifetime warranty. Maybe a historical dataset of production, or project management data (such as hardness of rock for tunnel boring machines), would be useful for training parametric linear relationships in a fuzzy TSK modeling and control approach.

Fuzzy controllers have enhanced performance even with multi-objective constraints, allowing conflicting conditions, because the fuzzy output is made to have the best compromise in the overall control strategy. For example, if a crane elevator is combined with a conveyor belt, or a rolling thread mechanism, and such a system takes very expensive loads from a container in a port, where the ship must stay as little as possible to pay less fees, the load must be transferred from the container to a dock as fast as possible, with no swinging of the load to avoid accidents, and still using the lowest energy and power as possible, with several safety alarms, a multi-objective controller can observe the distance from the ship to the container to the dock, the angle of the ropes that tie the load, and the power of the electrical motor to drive the load from the initial position until finally placed in a secure location. When a human operator controls this system, she or he will look into the load swinging and balance, and define by the desired transfer velocity what is the best angle of the ropes (or chains), at the same time looking into the signals of motor overload, anything crossing the traveling zone, any indication that the container is locked or not available, the destination dock place is not empty or has any issue, maybe it is wet, or it is raining. A human operator observes all, at the same time, and processes in parallel in his or her brain the best control action. It is pretty much impossible to design a PID control, with single-input/single-output for such multi-objective purposes. Another example would be in plastic injection molding, where the output temperature of the material, the quality of the molten raw materials, the pressure of the screw that will tight the manufacturing device, the environmental temperature, all of them are interrelated, and the flow of the resin will be dependent on several parameters and even the output of the system. Such a plastic injection molding would never have any PID control optimized for their operation, and a fuzzy logic system would be easily well adapted, a neural network controller could be trained based on past historical data. Artificial intelligence (AI)-based control would be the best choice for these kinds of ill-defined-modeling real-life industrial processes.

Conventional PID controllers and fuzzy controller can be sometimes combined, for example, Japanese kitchen appliances have had since 1970's commercial temperature with a PID for an extremely fast response, and just before the overshoot starts ringing, a fuzzy controller can take the command of the loop, imposing a flat response, exactly at the reference set point. It is expected that real-time operating systems will embed AI algorithms, and fuzzy controllers have a direct and simple way to be implemented. In the 1990s, there were some hardware implementations of fuzzy logic systems using Forth that is a language which easily takes new commands, compiles and incorporates in the running compiled code. MATLAB® has been very successful with their fuzzy logic toolbox, and neural networks toolbox, but to make a MATLAB live and compatible with a hardware implementation, it is necessary to have a lot of investments. Python is very slow, although it is very powerful; other recent languages that would allow a running fuzzy logic operating system would be Julia, Rust, and Swift.

5.2 Fuzzy controller heuristics

A conventional methodology, where the plant or process will have a mathematical identification, would need characterization with differential equations. A designer will need to define an analytical transfer function and synthesize a closed-loop control for a desired behavior, typically considering for a step response, the maximum overshoot, damping, velocity response, settling time, and steady-state error. However, assuming that the operator or control action is modeled, as presented in Figure 5.2, the procedure presented in Figure 5.3 can be implemented. Such a block diagram shows that a fuzzification module converts the crisp values of the control inputs into fuzzy values, so that they are compatible with the fuzzy set representation in the rule base. The choice of fuzzification strategy is dependent on the inference engine, i.e., whether it is composition based or individual-rule-firing based. The knowledge base consists of a database of the plant. It provides all the necessary definitions for the fuzzification process such as membership functions, fuzzy set representation of the input–output variables, and the mapping functions between the physical and fuzzy domains. The rule base is essentially for the control strategy of the system. It is usually obtained from expert knowledge or heuristics and expressed as a set of IF-THEN rules. The rules are based on the fuzzy inference concept, and the antecedents and consequents are associated with linguistic variables. There are two main types of fuzzy rule bases: (i) Type 1, or Mamdani, where the consequent (THEN) is a fuzzy set, or (ii) Type 2, or TSK, also called parametric or relational, where the consequent (THEN) has a linear equation that combines the input variables

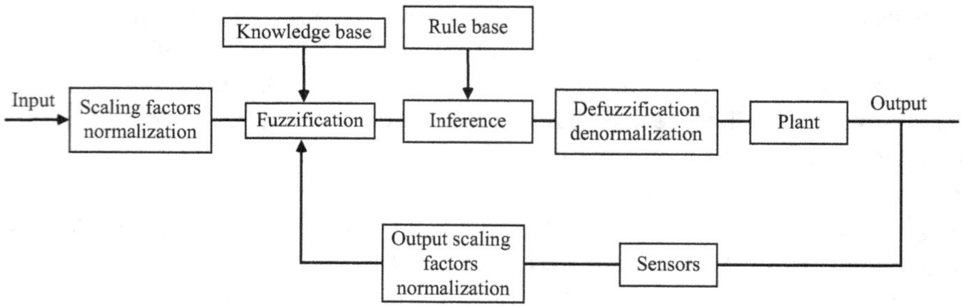

Figure 5.3 Block diagram of a typical fuzzy logic controller

in a multi-linear algebraic equation. The Type 2 fuzzy rule base is a hybrid of partially evaluating conditions of the input variables under a fuzzy logic framework mixed with a multi-linear equation modeling, which can be found with algebraic multi-linear regression. In the 1990s, there were presented and published the very first publications in fuzzy neural networks for power electronics (Simões and Bose, 1995, 1996a,b), such a technique is currently rebranded as a mixed fuzzy and neural network systems, called ANFIS (artificial neural fuzzy inference system), as in contemporary papers, books and available as a fuzzy logic toolbox of MATLAB.

The main difference between the Type 1 and the Type 2 techniques are (i) Type 1 serves for expert-based understanding of the system to be modeled or controlled, based on past experience and linguistic descriptions, while (ii) Type 2 serves for system with a lot of numerical data supporting a large database of past measurements of a system. Although Type 2 is very powerful and more precise in numerical computation, the need of data makes this approach less competitive when compared to neural network systems, because ANN can learn and train with input/output numerical data. In some cases, a Type 2 could be implemented as control or as model, but for most sophisticated and complex problems, it is advised to use a neural network to capture the system with some kind of feedforward learning topology. Neural networks can be used for input–output algebraic mapping, classification of patterns, and data compression, and in the past 15 years, a third wave of utilization of neural networks has been very successful and rebranded as "deep learning."

Figure 5.3 shows that after the inference engine is processed, it is necessary to do defuzzification, such as an inverse transformation, which maps the output from the fuzzy domain back into the crisp domain, as described in Chapter 4 of this book. In order to design a good fuzzy controller, we must have a good understanding of the physics of the process that we are trying to control.

Then we should write the rules, i.e., transfer our knowledge of how to properly drive the plant dynamics with the fuzzy controller. As an example, we can simply imagine an inverted pendulum, as depicted in Figure 5.4. The inverted pendulum is considered a very difficult nonlinear system for designing a controller using

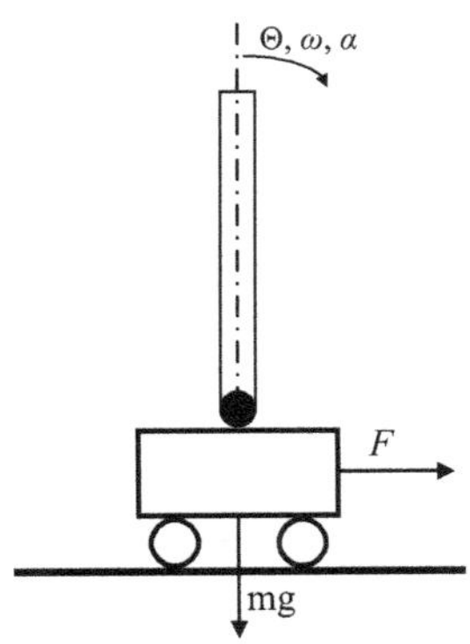

Figure 5.4 Definition of variables relevant to modeling an inverted pendulum

mathematical procedures. It is intrinsically unstable, because only keeping a driving force will maintain it upright, and lack of power will make it fall down because of gravity. Seven rules fuzzy system have been developed, as indicated in Figure 5.5, where experts helped to describe how to command a controller output as an action of keeping such an inverted pendulum balanced in their upright position, observing the angle and the derivative of the angle (angular velocity), from the vertical position. Those seven rules can be implemented based on fuzzification of input variables, and after the inference engine, the fuzzy set can be defuzzified using the center of gravity for the crispy control action, commanding the force to move the cart to the right or to the left. Of course, a real-life three-dimensional problem would be a way more complicated but still feasible to implement by adding more input data to be fuzzified with additional rules for generating a controller output, probably two motors would have to be constructed, moving the cart along a plane, instead of a line, as this is a very simplified theoretical case.

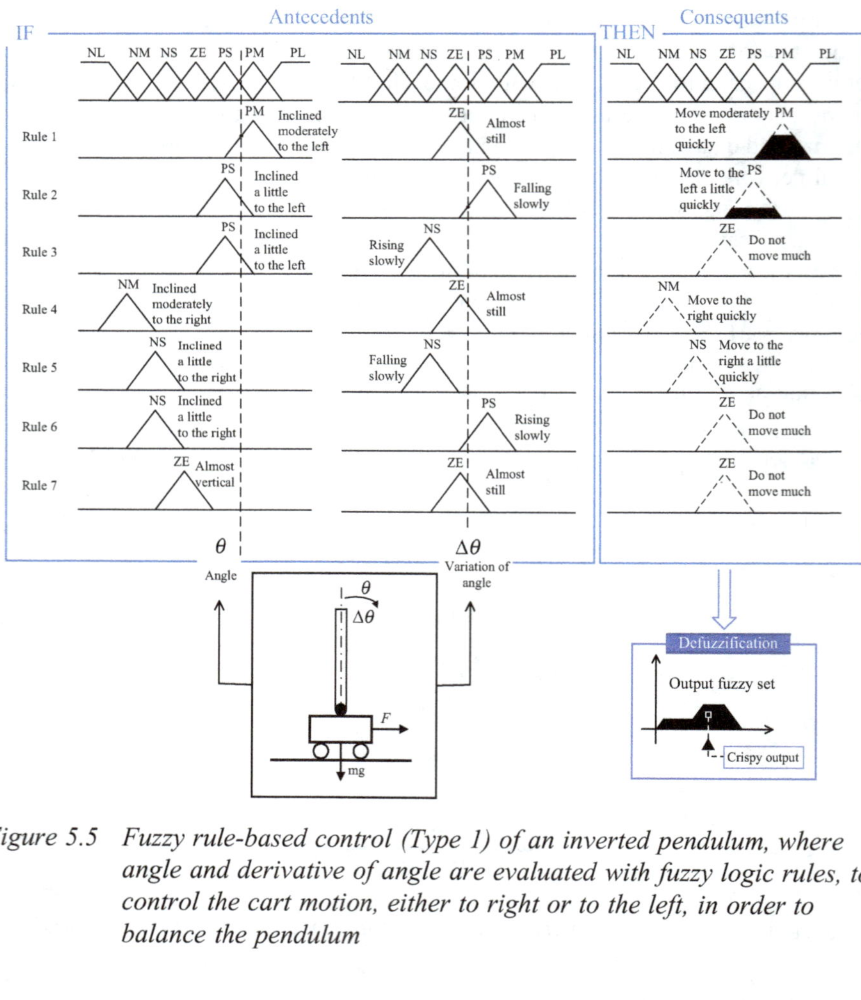

Figure 5.5 Fuzzy rule-based control (Type 1) of an inverted pendulum, where angle and derivative of angle are evaluated with fuzzy logic rules, to control the cart motion, either to right or to the left, in order to balance the pendulum

5.3 Fuzzy logic controller design

Conventional control has provided several methods for designing controllers for dynamic systems. All of them require a mathematical formulation for the system to be controlled and a certain approach that will be used in order to design a closed-loop control. Some of those methods are as follows:

- PID control: Over 90% of the controllers in operation today are PID controllers (or at least some form of PID controller like a P or PI or an I+P controller). This approach is often viewed as simple, reliable, and easy to understand. Sometimes fuzzy controllers are used to replace PID, but it is not clear yet if there are real advantages.
- Classical control: Lead-lag compensation, Bode and Nyquist methods, and root-locus design.
- State-space methods: State feedback, observers.
- Optimal control: Linear quadratic regulator, use of Pontryagin's minimum principle, or dynamic programming.
- Robust control: H2 or H∞ methods, quantitative feedback theory, and loop shaping.
- Nonlinear methods: Feedback linearization, Lyapunov redesign, sliding mode control, and backstepping.
- Adaptive control: Model reference adaptive control, self-tuning regulators, and nonlinear adaptive control.
- Stochastic control: Minimum variance control, linear-quadratic-Gaussian control, and stochastic adaptive control.
- Discrete event systems: Petri nets, supervisory control, and infinitesimal perturbation analysis.

These control approaches will have a variety of ways to utilize information from mathematical models. Sometimes they do not consider certain heuristic information early in the design process but use heuristics when the controller is implemented to self-tune (tuning is invariably needed since the model used for the controller development is not perfectly accurate). Unfortunately, when using some approaches to conventional control, some engineers become somewhat removed from the control problem itself and become more involved in the mathematics, this usually allows the development of unrealistic control laws. Sometimes in conventional control, useful heuristics are ignored because they do not fit into the proper mathematical framework, and this can cause problems. Fuzzy logic and neural networks approach exactly this lack of real-life understanding by heavily math-oriented control designers, by allowing heuristics and learning from past case studies or numerical data, retrofitting usually an excellent performance controller that most of the time excels when compared to heavy mathematical control design approaches.

An induction machine has a very complex instantaneous model based on decoupled *d–q* equations, based on trigonometrical Park and Clarke transformations. The *d–q* inverse machine model is resolved mathematically in a very small time step (about

100 μs), in order to control torque and flux with virtual *d–q* currents. Then, such set points are reversely calculated in real time in order to generate the pulse-width modulation of transistors in a three-phase inverter that commands the induction machine. An induction machine can also operate as an induction generator for a wind turbine. References (Simões *et al.*, 1997b; Souza *et al.*, 1997) are meant for journal publications of earlier presentations in IEEE IAS annual conferences, where the authors published for the first time how a double-PWM back-to-back converter can be controlled in real time, using a Texas Instruments DSP platform, to successfully implement a hardware controller for a vertical axis wind turbine system, to start up in motoring mode, capture the wind energy, implementing three fuzzy controllers: (i) FLC1, a maximum peak-power-tracking for optimizing the turbine aerodynamics efficiency; (ii) FLC2, a search algorithm to decrease the machine flux to improve generator core and copper losses; and (iii) FLC3, a fuzzy speed control to maintain the machine operating at the angular speed calculated by FLC1, which must be stable, resilient, parameter-insensitive and adaptive against wind vortex, gearbox/machine vibration, and turbine intrinsic pulsating torque. Simões *et al.* (1997b) and Souza *et al.* (1997) describe the real-time management for such a system to search the best induction generator velocity, locking it on peak-power-tracking, for then searching the best induction generator magnetic flux, locking it on improved efficiency, keeping a stable fuzzy-based wind control, with injection of active power in a three-phase utility grid.

The fuzzy logic control described by Simões *et al.* (1997b) and Souza *et al.* (1997) can also be implemented for other processes or industrial plants, and it has been extensively discussed in the literature for many other applications. It is based on a closed-loop control that calculates the instantaneous error (set point minus the feedback variable), executed in a discrete algorithm that saves the current error subtracted from the previous error in a variable called change-in-error. Those are the two inputs of the fuzzy control using Mamdani's inference engine, as depicted in Figure 5.6, where the output is considered to be variation-in-torque, scaled back from P.U. to rated value, integrated, forming the instantaneous torque set point to be used in the machine vector control

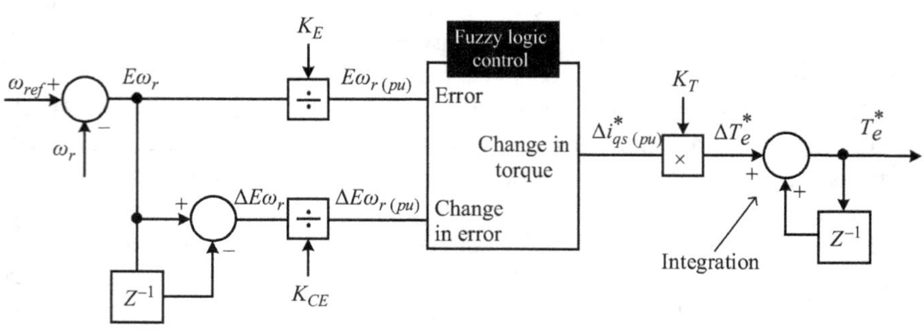

Figure 5.6 Fuzzy angular speed controller where error and change-in-error are scaled to P.U. (per-unit) to input a Mamdani's rule-based control, where the output is considered to be variation-in-torque, then scaled back to a rated value and integrated to have the instantaneous torque set point

scheme. Any general fuzzy logic control can be designed for an induction motor or DC motor drive or any process that would work with a PI or a PID control but may have parameter sensitivity, noise, and disturbance, making such an improved AI-based controller important to implement instead of a traditional PI controller. Figure 5.7 shows that the input signals for the fuzzy logic control are E (error) and CE (change-in-error) and the output (fuzzy rule table cell) is the derivative of output control, also called change-in-output, or variation-in-output. Figure 5.7 shows the fuzzy sets and their corresponding membership functions; fuzzy sets are linguistically defined in Table 5.1.

The universe of discourse is expressed in per-unit, and all membership functions are defined on such a range from -1 to $+1$. Therefore, it requires normalization. Figure 5.6 shows that the real-life error and change-in-error are multiplied by a scaling gain K_E and K_{CE}, respectively; therefore, the controller must be fine-tuned,

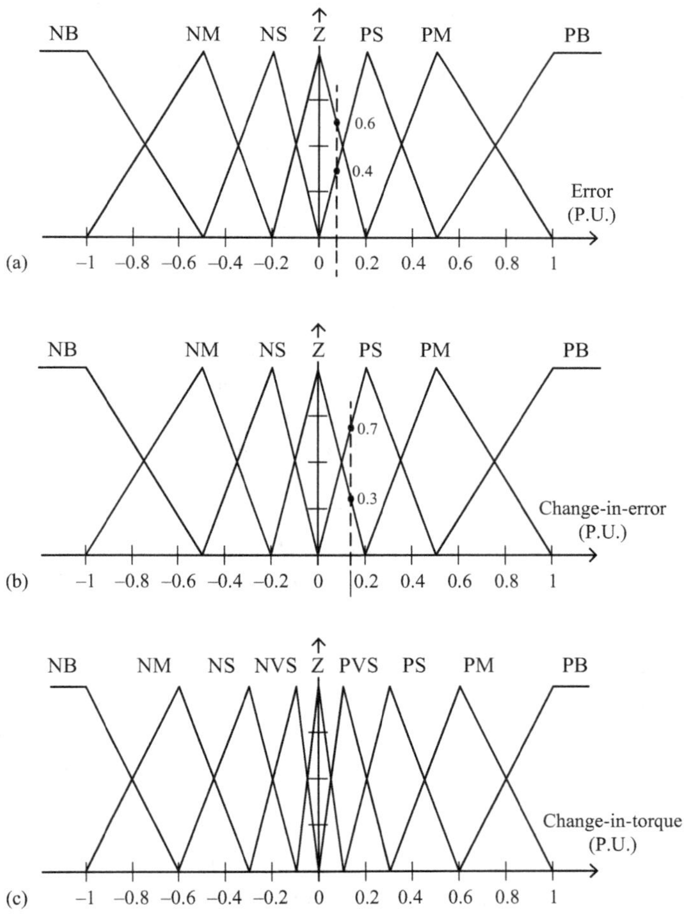

Figure 5.7 Fuzzy logic control membership functions with their associate linguistic variables: (a) error, (b) change-in-error, and (c) change-in-output

Table 5.1 Fuzzy linguistic variables for fuzzy control

NB	Negative big
NM	Negative medium
NS	Negative small
NVS	Negative very small
ZE	Zero
PVS	Positive very small
PS	Positive small
PM	Positive medium
PB	Positive big

Table 5.2 Fuzzy controller for a motor drive speed control loop

		Error E_{pu}						
		NB	**NM**	**NS**	**ZE**	**PS**	**PM**	**PB**
Change-in-error CE_{pu}	NB	NVB	NVB	NVB	NB	NM	NS	ZE
	NM	NVB	NVB	NB	NM	NS	ZE	PS
	NS	NVB	NB	NM	NVS	ZE	PS	PM
	ZE	NB	NM	NVS	ZE	PVS	PM	PB
	PS	NM	NS	ZE	PVS	PM	PB	PVB
	PM	NS	ZE	PS	PM	PB	PVB	PVB
	PB	ZE	PS	PM	PB	PVB	PVB	PVB

and in all possible transient conditions of the system, the normalized error and change-in-error should fit in a $[-1,+1]$ domain, for being fuzzified. Figure 5.7 shows that seven membership functions for each E_{pu} and CE_{pu} make a total of 49 possible rules. The output DU_{pu} is considered to have nine membership functions. Table 5.2 shows the fuzzy rule table, i.e., for each combination, for every fuzzy set of input signals E_{pu} and CE_{pu}, there is an output variable, considered to be a change-of-output DU_{pu}. The fuzzy rule in Table 5.2 shows that in each cell there is a consequent fuzzy set of a rule, assuming that there is an AND operation of those two inputs.

As discussed earlier, the rule matrix and membership functions of the variables are associated with the heuristics of general control rule operation, i.e., the meta-rules, such heuristics would be the same way an expert would try to control the system if an operator would be in the feedback control loop themselves. The rules are all valid in a normalized universe of discourse, i.e., the variables are in per-unit. For a simulation-based system design, the controller tuning can be done with the fuzzy logic toolbox of MATLAB, and LabView is another nice environment for such design. It is also possible to develop the whole structure of the controller using C language compiled code. For advanced design, it is possible to use neural network or genetic algorithm techniques for fine-tuning the membership functions, implementing an adaptive neuro-fuzzy inference system (ANFIS). Such details are outside the scope of this chapter. This fuzzy speed control algorithm can be numerically explained and clarified with the following step-by-step procedure:

1. Sample set-point speed ω_r^* and actual shaft-speed ω_r.
2. Computer error, change-in-error, and then per-unit values are as follows:

$$E(k) = \omega_r^* - \omega_r$$

$$CE(k) = E(k) - E(k-1)$$

$$E_{pu} = \frac{E(k)}{GE} = K_E E(k)$$

$$CE_{pu} = \frac{CE(k)}{GCE} = K_{CE} CE(k)$$

3. Suppose the fuzzy logic control is written in compiled language. Therefore, the commands are sequentially executed. This step 3 must identify in the universe of discourse where such variables, error and change-in-error, are located. Two indexes I and J for E_{pu} and CE_{pu} can be used to define which fuzzy sets are triggered by the sampling of the error and the change-in-error. If the membership functions are of simple triangular or trapezoidal shapes, each index I and J will help us to decide a linear equation to be used to calculate the membership function evaluation in the next step.
4. Calculate the degree of membership for each fuzzy set for E_{pu} and CE_{pu} by applying the correct equations identified in step 3, defining exactly which rules in the rule table will be used (usually four are triggered, assuming full overlapping of membership functions).
5. Identify the four valid rules; all the rules are stored in a chain of IF–THEN–ELSE statements, or maybe in a lookup table, or with a similar functional implementation. It is necessary to calculate the individual inference engine rules' contribution, called rule degree of validity, rule degree of strength, rule firing value, or rule degree of truth. Normally, AND operator is used, which can be implemented either as MIN of two membership functions, or by the multiplication of the values of membership functions of the input variables. In this step, four rules will have each their degree-of-truth, to be used next for making up the fuzzy output and then calculate the defuzzification.
6. Compose the output fuzzy set and get it prepared for defuzzification by the center of gravity, or use a simplified height defuzzification method, which is a simple weighted average of the output fuzzy singletons weighed by the degree-of-truth of each rule.
7. Since the output is a change or a derivative, an integration, or accumulative sum, must be performed. For the fuzzy rule table of a PI controller, it is necessary to perform such a numerical integration, because the output of the rule base is change-in-output (variation in torque). There are some other rule tables developed for the full output (not variation), and then, such integration is not necessary.

For example, a typical rule in the matrix is like Rule 30 that reads as

$$\textbf{\textit{IF}}\ E_{pu} = PS\ \textbf{\textit{AND}}\ CE_{pu} = PM\ \textbf{\textit{THEN}}\ DU_{pu} = PB \tag{5.1}$$

The general considerations in designing such a fuzzy controller are expressed by the following meta-rules:

- If both E_{pu} and CE_{pu} are zero, maintain the present control setting, i.e., $DU_{pu} = 0$.
- If E_{pu} is not zero but it is approaching at a good rate, i.e., CE_{pu} has a good polarity and value then maintains the present control setting, i.e., $DU_{pu} = 0$.
- If E_{pu} increases, it changes the control signal DU_{pu} to bring the plant output back to the conditions where error should be zero, i.e., reverses the trend and makes signal DU_{pu} dependent on the magnitude and sign of E_{pu} and CE_{pu} in order to force E_{pu} toward zero.

In several other fuzzy logic control systems, some rules do not show, either because it was found that such situations do not exist, or maybe because the output should not change. Table 5.2 depicts a fuzzy logic controller with 49 possible combinations that can be used for a speed control of motor drive. The fuzzy rule table has the derivative (change or variation) of the output; for induction motor drive, it is considered the q—torque-current Δi_{qs}^{*} in per-unit, which is multiplied by a scaling factor (K_T) to normalize to the change of torque of the motor; such a variation of torque must be accumulated (or integrated) in order to build up the torque control command. When the output control signal is integrated, as indicated in Figure 5.6, this rule table operates as fuzzy PI control. There are some applications of the same rule table, where the same output signal of the same table is assumed to be a total full output, instead of a change-in-output. Therefore, the rule would generate an output signal, without any integration, and for such structure the controller becomes fuzzy PD control. The technique presented in the chapter for fuzzy PI control mostly works really well, as long as good scaling factors are retrofit for the particular application. It would be important to have some simulation studies and some trial-and-error tweaking on the controller. The implementation can be done in a fixed-point DSP or microcontroller, with proper integer calculations. If the target hardware is floating-point, the simulation can be translated with the same scaling factors and per-unit parametrization. For real-time control, the inference engine can be simplified for MIN operation on the AND input fuzzy sets, generating a degree-of-truth for each rule (or also called rule-output strengths). The output could be calculated on center of gravity, but in PI closed-loop control, the system will stabilize to zero-steady-state error, so it is easier to assume that fuzzy outputs are singletons, instead of the whole fuzzy membership functions, by considering only where the peak of each fuzzy output set occurs, and a height defuzzification, which is a balanced weighted average, will convert the fuzzy inference into a crispy real value, i.e., when "singletons" are used to simplify the defuzzification process, the output fuzzy set becomes a membership function represented by a single vertical line, as discussed in Chapter 4. Then, the center-of-gravity (or center-of-area) calculation would reduce to just a weighted average calculation of numbers and rule strengths, with the rule strengths used as weights, making it very easy to implement in any microprocessor, microcontroller, DSP, or RISC-based hardware. In order to implement such a controller in a compiled language, such as C, C++, Forth, and Rust, the following data manipulation and numerical calculation must be accommodated in a real-time control loop:

- System inputs
- Input membership functions
- Antecedent values
- Rules
- Rule-output strengths
- Output membership functions
- System outputs

5.4 Industrial fuzzy control supervision and scheduling of conventional controllers

Most controllers in operation today have been developed using conventional control methods. There are many situations where these controllers are not properly tuned, and there is heuristic knowledge available on how to tune them while they are in operation. Fuzzy control methods can be used as the supervisor that tunes or coordinates the application of conventional controllers. Figure 5.8 illustrates how an outer loop might be implemented in supervising a multiloop PID industrial process. Each PID controller is best tuned with industrial practices. Experts and industrial process engineers would help in defining their operation and construct a database for a fuzzy inference engine, maybe partially rule based (Type 1) or also parametric based (Type 2), when set points for those PIDs would be readjusted in accordance with such a hybrid fuzzy-based supervisory manager. The Type 2 equations could define set points based on multi-linear regression of supply and demand operating prices, electricity and thermal energy costs, allocated human resources for specific shifts or holiday off-seasons, and any econometric-based modeling that would help a process management engineer to set the operating points of such an industrial process. Most of the controllers in operation today have been developed using conventional control methods. There are many situations where these controllers are not properly tuned, and there is heuristic knowledge available on how to tune them while they are in operation. Fuzzy control methods

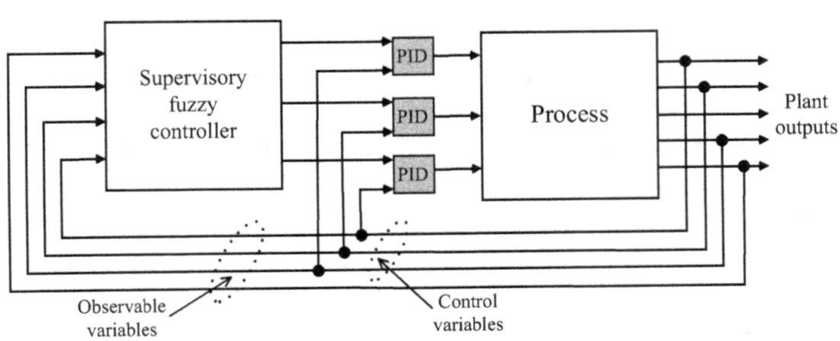

Figure 5.8 Industrial fuzzy supervisory management control

can be used as the supervisor that tunes or coordinates the application of conventional controllers.

Almost the majority of the controllers in operation are PID controllers; industrial and control engineers have been applying simple procedures for designing them, and they are often available at little extra cost because they can be incorporated into the programmable logic controllers that are used to control many industrial processes. As explained in the supervisory control discussion, many of the PID loops that are in operation are in continual need of monitoring and adjustment since they easily become improperly tuned. While there exist many conventional methods for PID auto-tuning, it would be possible to design a supervisor trying to recognize when the controller detuned, then seeking to adjust the PID gains to improve performance. Therefore, a "behavior recognizer" will seek to characterize the current behavior of the plant in a way that will be similar to an indirect adaptive controller. Simple tuning rules may be used where the premises of the rules form part of the behavior recognizer and the consequents form the PID designer, some possible fuzzy rules are as follows:

IF the steady-state error is *LARGE* **THEN** increase the proportional gain.
IF the response is *OSCILLATORY* **THEN** increase the derivative gain.
IF the response is *SLUGGISH* **THEN** increase the proportional gain.
IF the steady-state error is *TOO BIG* **THEN** adjust/down the integral gain to decrease error.
IF the overshoot is *TOO BIG* **THEN** decrease the proportional gain.

Some plants or processes are characterized by common nonlinearity associated with a slow thermal process or saturation in the magnetic core of transformers, inductors, and electrical machines, i.e., depending on the temperature the transfer function may have either increasing or decreasing gain, and a PI controller adjusted for the middle range of such nonlinearity would have a different steady-state error. Suppose a heat exchanger is characterized by a nonlinear control law with an augmented piecewise linearization as depicted in Figure 5.9. If three regular PI controllers can be optimized for each center of every region, three fuzzy TSK rules can be implemented:

$$\boldsymbol{IF} E_{pu} = N \ \boldsymbol{THEN} \ DU_{pu} = a_{10} + a_{11} \times E + a_{12} \times \Delta E$$

$$\boldsymbol{IF} E_{pu} = Z \ \boldsymbol{THEN} \ DU_{pu} = a_{20} + a_{21} \times E + a_{22} \times \Delta E$$

$$\boldsymbol{IF} E_{pu} = P \ \boldsymbol{THEN} \ DU_{pu} = a_{30} + a_{31} \times E + a_{32} \times \Delta E$$

The coefficients a_{ij} are proportional, integral, and derivative gains of three optimized PIDs, the output of this controller DU_{pu} must still be integrated, otherwise the formulation makes it to be a PI controller with an offset. This simple fuzzy PI has scheduled gains for a nonlinear system, which can also be made adaptive by incorporating a recursive least-square time-window for learning the best coefficients a_{ij}.

Experts and industrial process engineers would help in defining their operation and construct a database for a fuzzy inference engine, maybe partially rule based (Type 1) or also parametric based (Type 2), when set points for those PIDs would be readjusted in accordance with such a hybrid fuzzy-based supervisory manager.

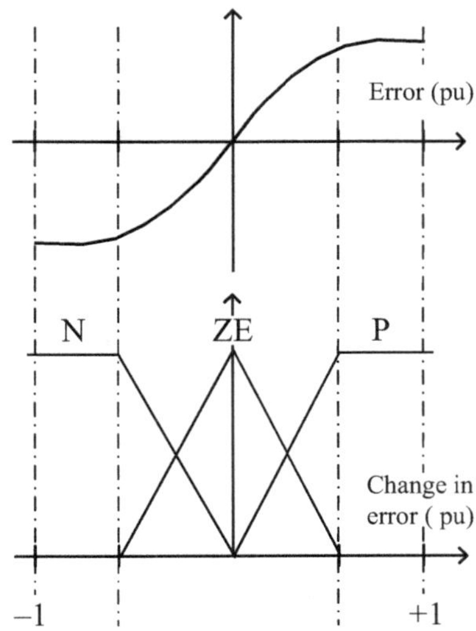

Figure 5.9 Nonlinear heat-exchange control law approached by three linear segments approximate by three input fuzzy sets for three parametric linear equations of TSK inference

The Type 2 equations could define set points based on multi-linear regression of supply and demand operating prices, electricity and thermal energy costs, allocated human resources for specific shifts or holiday off-seasons, and any econometric-based modeling that would help a process management engineer to set the operating points of such an industrial process.

Chapter 6

Feedforward neural networks

The field of neural networks (NNs) has a history starting just after the WW II, with sounding industrial applications in the past 40+ years, more recently in deep learning paradigms. Chapter 7 has a history timeline portrayed in Figure 7.8, showing that the backpropagation algorithm revitalized the field of NNs in 1985 and became a solid training technique. The field of NN became suffused with applications by the end of the 1990s, on a diversity of paradigms, and learning methods. Many successful approaches have been categorized as their paradigm of NN topology versus their industrial applications (Meireles *et al.*, 2003).

Internet companies allowed to develop massive data applications for audio and video streaming, social networking, live video broadcasting, peer-to-peer and group communications, channels for workgroups. There has been a worldwide interaction of people with diverse interests and a need in developing mathematical models for supporting decisions on such massive data. NNs would be a natural choice, because they divide-and-conquer: each neuron in a network can solve one small part of a larger problem, so that the overall problem would be solved by combining these solutions.

By the end of the first decade of the twenty first century, there was a third rebirth of the field of NNs, with the paradigm rebranded as deep learning. As regarded to electrical power systems and smart-grid applications, further details are discussed in Chapter 9 of this book.

The availability of computers with the power to perform simulations and the development of specialized hardware to implement NNs helped the expanding interest and research in NNs. NN technology is being applied to solve a wide variety of scientific, engineering, and business problems, and to perform complex functions such as noise cancellation, adaptive filtering, pattern recognition, nonlinear controls, and econometric forecasting. There are four main characteristics that make NNs so valuable:

- They can learn relationships between input and output data. Such learning does not depend on the programmer's prior knowledge of rules. They can infer solutions from presented data, often capturing subtle relationships.
- NNs can generalize and handle noisy, imperfect, or incomplete data. Such generalization features provide a measure of fault tolerance and is useful when examining real-world data.
- They can capture complex, higher order functions, and nonlinear interactions among the input variables in a system.

- NNs are highly parallel; their numerous operations can be executed simultaneously in most of the topologies. Parallel hardware can execute hundreds or thousands of times faster than conventional microprocessors, making many applications practical for the first time.

The development of NNs was inspired by the studies for understanding the biological nervous system. Preliminary theoretical foundations on physiology and psychology for neural networks were proposed by Alexander Bain (1873) and William James (1890). In their work, both thoughts and body activity resulted from interactions among neurons within the brain. Their concepts foretold the notions of a neuron's activity as being a function of the sum of its inputs. Half a century later, McCulloch and Pitts (1990) published a seminal paper, in which they derived theorems related to models of neuronal systems based on what was known about biological structures in the early 1940s, showing that a network could represent any finite logical expression with a massively parallel architecture. In 1949, Hebb (1949) published a book, where he defined a method to update synaptic weights for what is now referred to as Hebbian learning. The landmark by Rosenblatt (1962) defined an NN structure called perceptron. It was simulated in detail on an IBM 704 computer at the Cornell Aeronautical Laboratory and caught the attention of engineers and physicists because such a computer-oriented paper described the perceptron as a "learning machine." This paper laid the groundwork for both supervised and unsupervised training algorithms as they are today in backpropagation and Kohonen networks, respectively.

In 1960, Widrow and Hoff (1960) published a paper where they had simulated NNs in computers and also had implemented their designs in hardware. They introduced a device called an ADALINE, an adaptive linear processing unit based on a neuron. An ADALINE consists of a single neurode with an arbitrary number of input elements that can take on values of plus or minus one and a bias element. Before being summed by the neuron-summer circuit, each input (including the bias) is modified by a gain. The Widrow–Hoff algorithm is a form of supervised learning that adjusts the weights according to the error intensity at the output of the summer. They have shown that their technique for adjusting the weights could minimize the sum-squared error over all patterns in the training set.

The first wave of artificial NNs (ANNs) research can be associated with the frequency of the keyword "cybernetics," which started around 1940s and peaked at 1970s. The development of NNs slowed at the end of the 1960s and middle of the 1970s, mainly because Minsky and Papert (1969) cooled off the NN community with a book called *Perceptrons*. In such a book it was presented a comprehensive analysis on a single perceptron, without hidden layers. However, their writing style was full of criticism in claiming that most of the research about NNs was "without scientific value." They showed that the two-layer perceptron was rather limited, because it could only work with problems associated with linear separable solution spaces. The Exclusive-OR (XOR) problem was used on an elementary system that the perceptron was unable to solve. About the year 1969 people only knew how to train two-layer networks; there was no effective algorithm to train a network with three or more layers. The derivation of the backpropagation algorithm came later, contemporary to the ADALINE. It returns the value of a function to learn and predict from data (Widrow and Hoff, 1960), while the training algorithm to adapt the weights of the ADALINE was a stochastic gradient descent, and slightly different versions are still used today. After the book by Marvin Minsky, virtually all research funds for NN dried up.

In the 1980s the second wave of NN research emerged thanks to connectionism, based on cognitive science and symbolic reasoning, trying to explain and associate how the brain works and implement the idea of biological nervous systems as hidden units in computational models. Associated to connectionism there was also the approach of parallel distributed processing (Goodfellow *et al.*, 2016). Particularly important to revive the field was the research conducted by Hopfield (1982) in the beginning of the 1980s, leading to today's Hopfield network. However, what primarily influenced people about the capabilities of NNs was the discovery of the backpropagation algorithm. It was first found by Werbos (1974) and Parker (1982), and independently rediscovered and popularized around 1985 by Rumelhart *et al.* (1986).

Backpropagation is a gradient descent system that tries to minimize a cost function (usually the mean squared error (MSE)) by moving down the gradient of the error curve. This algorithm updates the weights by calculating local errors for each neuron, allowing the use of hidden layers in the NN topology. Although there are approximately more than 100 network topologies and several other training methods, a feedforward NN can be trained by regular backpropagation method. It works well for one hidden layer, sometimes two hidden layers, but the training usually does not converge for more than two or too many neurons in two hidden layers. Such a practical observation made designers to avoid "deep networks" for a long time, making such a second wave of NNs research to last until the end of the 1990s.

Other fields of machine learning made advances at the same time, such as kernel support vector machines, graphical models, and powerful computers allowed Bayesian statistics and multidimensional algebraic solutions to be implemented to increasing dataset sizes and increasing model sizes, slowing down the use of connectionism and declining the popularity of NNs to a trailing edge about the year 2007. The connectionism ideas were realist but actually did not lead to any improvement in machine learning performance. The second wave of NNs lasted from 1985 to 2007, and it made possible the development of scores or hundreds of NN architectures, a breakthrough by Geoffrey Hinton when he presented in 2006 a deep-belief network efficiently trained using a strategy called greedy layer-wise pretraining (Goodfellow *et al.*, 2016). Therefore, the second rebirth and our current third wave of NNs research were triggered by the development of the rectified linear unit, or ReLU, and were presented in Goodfellow *et al.* (2016). The reader should refer to Chapter 9 in this book for further information.

6.1 Backpropagation algorithm

Despite the many descriptive names applied to the various paradigms, all NNs perform the same fundamental function: they map one vector space onto another, it can be an input/output mapping where input data and output data are used with a "teacher algorithm" to make the NN to learn the algebraic relations, or it can be a mapping of classes or clusters, where there is categorization. The NN could also learn the history of data and have an internal memory to associate a data series with a mapping of past events that can help one to forecast the sequence of next ones. A major accomplishment of the connectionist movement has been the successful use of backpropagation (BP) to train NNs (Goodfellow *et al.*, 2016). The BP algorithm waxed, waned, was criticized, adapted, but still is as a dominant approach, usually associated with other ones, in training modern deep learning NNs.

A feedforward NN topology has one input layer, one or more hidden layers, and one output layer. An input vector is applied to the input layer of the network, and in response, the network produces an output vector. Each vector consists of one or more components, each of which represents the value of some variable. Mapping may be static; in which case a feedforward NN will suffice. It could otherwise be dynamic, involving previous network states; in this case, a feedforward network may use delayed inputs or a network with feedback, such as a recurrent NN (although there are other recursive paradigms). The utilization of the hidden layer with nonlinear units allows the network to develop any kind of mapping, not just linearly separable ones.

A backpropagation network operates in two steps during training. First, an input pattern is presented to the network's input layer. The resulting activity flows through the network from layer to layer until the output is generated. Next, the network output is compared to the desired output for that input pattern. The error is passed backward through the network—from the output layer back to the input layer, with the weights being modified as the error backpropagates. A typical neuron has several inputs (x_j) that are multiplied by the correspondent weights (w_i) as shown in Figure 6.1. Each neuron computes the weighted input as (6.1). This summation is passed through the activation function, also called a "squashing" function. A typical activation function is usually smooth and nonlinear, like the sigmoidal function given by (6.2) depicted in Figure 6.2.

$$I = \sum_{j-1}^{n} w_i x_j \tag{6.1}$$

$$\phi(I) = \frac{1}{1 + e^{-I}} \tag{6.2}$$

The sigmoidal activation function has the convenient property that their derivative is a relationship of the same function, as indicated in (6.3). The weight

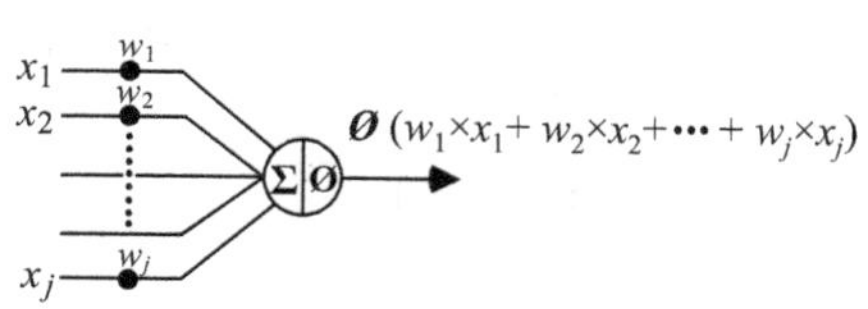

Figure 6.1 Structure of a neuron element

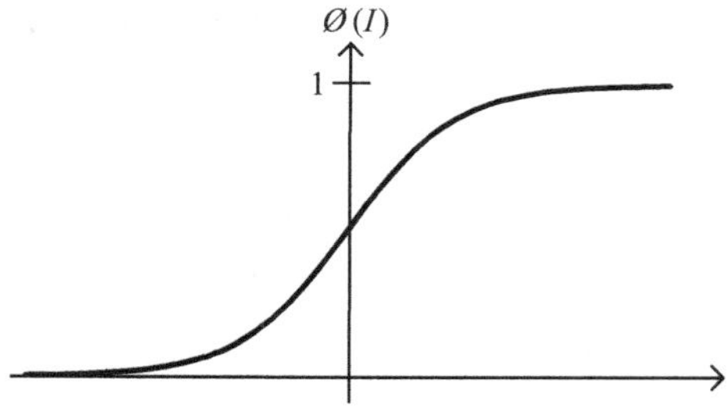

Figure 6.2 Sigmoidal activation function

change law is given by the generalized delta rule, where the change in a given connection weight is proportional to a learning coefficient (β) and to the partial derivative of the error function (E) in respect to the weight that is being modified in accordance with (6.4):

$$\frac{d\varnothing(I)}{dI} = \varnothing(I)[1 - \varnothing(I)] \tag{6.3}$$

$$\Delta\omega_{ij} = -\beta\frac{\partial E}{\partial\omega_{ij}} \tag{6.4}$$

The partial derivative of error in respect to each weight connection can be expanded by the chain rule $\Delta\omega_{ij} = -\beta(\partial E/\partial a_k)(\partial a_k/\partial net_k)(\partial net_k/\partial\omega_{ij})$; the hidden-to-output connections the previous gradient descent rule gives in the following equation:

$$\Delta\omega_{ij} = \beta E_j\varnothing(I) = \beta\left[y_j^{desired} - y_j^{actual}\right]\varnothing(I) \tag{6.5}$$

For the output layer the error that causes the weight change is evident as it is the difference between the network output and the desired one. But in the middle layer, a more complex procedure is followed. For the input-to-hidden connections it is necessary to differentiate with respect to the weights by applying the chain rule, as given by (6.6). For high-order error surfaces, the gradient descent method can be very slow if the learning coefficient β is small and can oscillate widely if β is too large; the addition of a momentum term can overcome such problem, indicated in (6.7). It provides to each connection some inertia, so that it tends to change in the direction of the average "downhill force," this scheme is implemented by giving a contribution from the previous weight change to each new weight revision:

$$\Delta w_{jk} = \beta\phi_k(I)[1 - \phi_k(I)]\sum_{k-1}^{n} w_{jk}E_k^{output}\phi_j(I) \tag{6.6}$$

$$\Delta w_{ij} = -\beta\frac{\partial E}{\partial w_{ij}} + \alpha\left(\Delta w_{ij}\right)_{previous} \tag{6.7}$$

6.2 Feedforward neural networks—a simple binary classifier

The *perceptron*, depicted in Figure 6.3, is considered as an "algorithm for learning a binary classifier," a function that maps a multi-input x_k, i.e., a multi-valued real-vector to an output value (a single output), which might be binary, "0" or "1," as indicated by the following equation:

$$f(x) = \begin{cases} 1, \text{if} \sum_{k=0}^{m} w_k \cdot x_k + b > 0 \\ 0, \text{otherwise} \end{cases} \tag{6.8}$$

Widrow and Rosenblatt were the first ones to study the perceptron. In such a context, a perceptron is one artificial neuron, where its output is binary. It uses the Heaviside step function as the activation function $f(\cdot)$. Therefore, the perceptron algorithm could be defined as a single-layer neuron. As a linear classifier, the single-layer perceptron is the simplest feedforward NN. Figure 6.3 shows that w is a vector of real-valued weights, $w \cdot x$ is the dot product, n is the number of inputs in the perceptron. There is an extra fixed input, called bias, that could be denoted as b. The bias signal makes the decision boundary away from the zero value (shifting), and it does not depend on any input value. The bias signal may also have a weight to be trained for such shifting.

The value of $f(x)$, to be either 0 or 1 as a binary output, can be used to classify the combination of inputs into two regions. If b is negative, then the weighted combination of inputs must produce a positive value greater than $|b|$ in order to push the classifier neuron over the 0 threshold. Spatially, the bias alters the position of the decision boundary. The perceptron learning algorithm will terminate the learning only if such a set is "linearly separable," i.e., there is a boundary that clusters two regions above and below a hyperplane. If the vectors are not linearly separable, learning will not reach such a point where all vectors are classified properly. The famous example of such inability in perceptron computing was discussed and extensively studied with a simple Boolean Exclusive-OR or Exclusive-NOR problem, which gave support to the book by Minsky and Papert (1969).

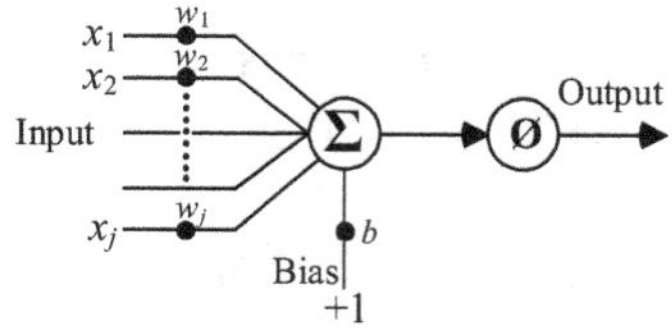

Figure 6.3 Perceptron, where several inputs are multiplied by a weight, summed and the output function can be linear, nonlinear, or compared to a threshold

AND

x_1	x_2	y
1	1	1
1	−1	−1
−1	1	−1
−1	−1	−1

OR

x_1	x_2	y
1	1	1
1	−1	1
−1	1	1
−1	−1	−1

XOR

x_1	x_2	y
1	1	−1
1	−1	1
−1	1	1
−1	−1	−1

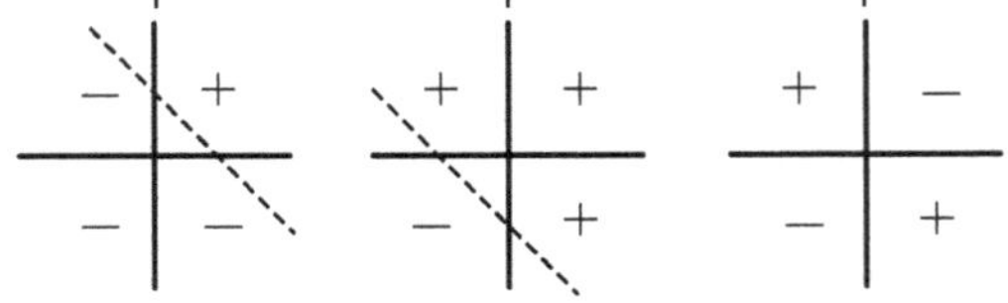

Figure 6.4 There is no straight line allowing a class separation for the XOR

When a training weight vector allows an output +1 for a group of inputs versus −1 otherwise, the problem is considered to be "linearly separable." Figure 6.4 shows how the functions AND and OR would have a single straight line, serving as the threshold, to separate two classes, while an XOR would not have such a straight line. Minsky showed that a single-layer network may only learn separable problems, and only multilayer networks can be trained for nonlinearly separable problems, i.e., a three-layer network with nonlinear activation functions can learn non-separable problems. Although there were several researchers that studied gradient-descent methods for training classifiers (Werbos, 1974; Parker, 1982; Rumelhart *et al.*, 1986; Amari, 1967), a team-work of Rumelhart *et al.* (1986) made possible a collection of papers in 1986, where they proposed a consistent method for training a multilayered feedforward NN.

The backpropagation algorithm became very popular, allowing the widespread use of NNs after 1986. Learning is the process by which the free parameters of an NN are adapted through a process of stimulation by the environment, and back-propagation is a prescribed set of well-defined rules for conducting an NN to learn. Consider one output neuron, for example, the one at the output layer of a multilayer perceptron (MLP), as indicated in Figure 6.5, such an output neuron will produce $y_m(k)$ that is compared to the desired value $d_m(k)$, and the error signal $e_m(k)$ must drive the incoming weights to be adjusted, in order to minimize such an error. The error signal actuates as a control mechanism to correct the synaptic weights. Such objective is achieved by minimizing a cost function, or index of performance, such as the following equation where $e(k)$ is the instantaneous value of the error energy:

$$e(k) = \frac{1}{2} \cdot e_m^2(k) \tag{6.9}$$

The minimization of $e(k)$ leads to a learning rule commonly referred to as the delta-rule or Widrow–Hoff rule (Widrow and Hoff, 1960). Let $w_{mn}(k)$ denote the synaptic weight of neuron m excited by element $x_n(k)$ of the signal vector $\tilde{x}(k)$ at the time step k. In accordance with the Widrow–Hoff rule, the adjustment $\Delta w_{mn}(k)$ is defined by (6.10), where η is the learning rate gain. The adjustment makes the synaptic weights of a neuron proportional to the product of the error signal and the

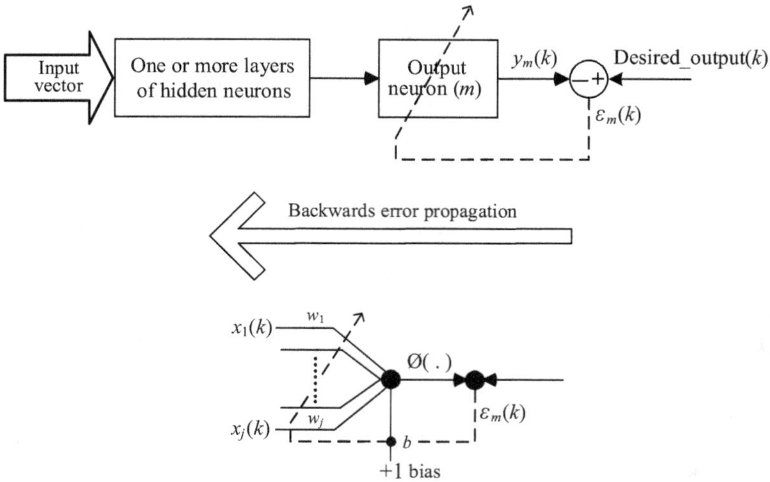

Figure 6.5 Correction of weights of connections to an output neuron, where the output is compared to the desired value, the error signal update weights in order to minimize such an error

input signal of the synapse in question. Having computed the synaptic adjustment $\Delta w_{mn}(k)$, the next weight value is given by the following equation:

$$\Delta w_{mn}(k) = \eta \cdot e_m(k) \cdot x_n(k) \tag{6.10}$$

$$w_{mn}(k+1) = w_{mn}(k) + \Delta w_{mn}(k) \tag{6.11}$$

6.3 Artificial neural network architecture—from the McCulloch–Pitts neuron to multilayer feedforward networks

Most real-world problems can be formulated as an equivalent formal mapping problem, or an algebraic computation of inputs that calculate outputs. Such a map can be represented by an ANN, and the building blocks are McCulloch–Pitts artificial neurons. The McCulloch–Pitts neuron was first proposed in 1943 by Warren McCulloch (a psychologist) and Walter Pitts (a mathematician) describing how simple artificial representations of neurons could represent any arithmetic function. In order to actually implement such a function, the psychologist Donald Hebb described in his book "The Organization of Behavior" in 1949, what today is called *Hebbian learning*: it is a reinforcement learning where right answers emphasize paths by increasing the weights associated with these paths. Eventually, reinforcement learning became again important in the current third wave of NNs research, i.e., there are deep learning structures based on RL applied to convolutional NNs and recurrent NNs based on the LSTM unit (described in Chapter 9).

The learning rule proposed by Hebb was the first mechanism for determining the weights of an NN based on how humans and animals learn. Hebb's rule can be used to train multilayered NNs. There are many different NN paradigms and many algorithms for determining the weights of an NN. Most of these algorithms work iteratively, i.e., starting with a random set of weight values, then applying one or more samples of the mapping, and gradually updating the weights. This iterative search for a proper weight set is called the learning or training phase.

Figure 6.6 depicts a general McCulloch–Pitts neuron that can easily be understood as a simple mathematical operator. It has several inputs and one output and perform two elementary operations on the inputs: first it takes a weighted sum of all the inputs, and then it applies a transfer function in order to send out the output. Such an artificial neuron can be written as a mathematical function, taking N inputs $\{x_1, x_2, \ldots, x_N\}$, or an input vector $\bar{x}$ and considering the output a scalar y. The output y can be expressed as a function of its inputs according to the following equations:

$$sum = \sum_{i=1}^{N} x_i \tag{6.12}$$

$$y = f(sum) \tag{6.13}$$

It is common practice to apply an appropriate scaling to the inputs (usually such that either $0 < x_i < 1$, or $-1 < x_i < 1$). Before summing the inputs, they have to be modified by multiplying them by a weight vector $\{w_1, w_2, \ldots, w_N\}$, so that the weighted sum of the inputs is calculated in accordance to with (6.14). The transfer function or the activation function $f(\cdot)$ could be just a threshold function, i.e., giving an output of 1 when the sum exceeds a certain value, and zero when the sum is below such value, or it can be a nonlinear transfer function. It is common practice to adopt the sigmoid function, which can be expressed as (6.15).

Training or learning paradigms for an ANN can be mainly classified as supervised or unsupervised. In supervised learning, inputs and targets (desired outputs) are known, and the ANN model is trained in a way that maps inputs to the outputs. Supervised learning is employed for regression and classification

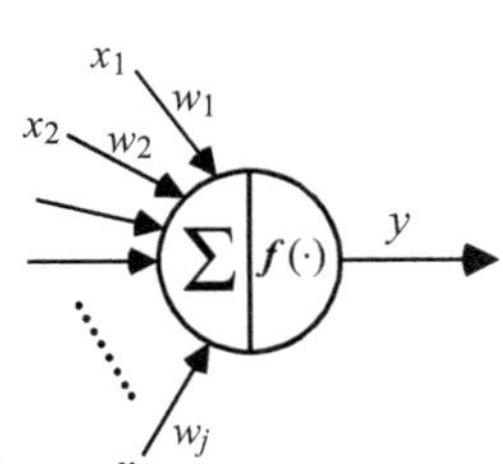

Figure 6.6 The McCulloch–Pitts artificial neuron as a mathematical operator

purposes. However, in unsupervised learning, targets are unknown, and the underlying relationship within datasets must be disclosed by the ANN using the data clustering method. Unsupervised learning can be used for filtering (de-noising), clustering of data, and finding deep patterns:

$$weighted\ sum = \sum_{i=1 \rightarrow N} w_i \cdot x_i \tag{6.14}$$

$$f(\cdot) = \frac{1}{1+e^{-(\cdot)}} \tag{6.15}$$

6.3.1 Example of backpropagation training

The procedure for the backpropagation algorithm is worked here as an example. Suppose there is a fully connected feedforward NN. The input i has two neurons, the middle layer j has two neurons, and the output layer k has one neuron. Assume the learning coefficient β is 0.7 and the network of Figure 6.7 shows all neurons with sigmoidal functions and unity slope gain. The initial weight values are randomized as indicated in Table 6.1.

Using weight updating (6.3)–(6.5), if the actual output is 0.7 and the target is 0.45, the weight update for the output layer in this pass will be

$$\Delta\omega_{01} = (0.7)(0.574)(0.45)(1\ -\ 0.45)(0.7\ -\ 0.45)\ = 0.024861$$

$$\omega_{01\ NEW} = 0.2\ +\ 0.024861\ =\ 0.2248$$

$$\Delta\omega_{02} = (0.7)(0.475)(0.45)(1\ -\ 0.45)(0.7\ -\ 0.45)\ = 0.02057$$

$$\omega_{02\ NEW} = -0.5\ +\ 0.02057\ =\ -0.479$$

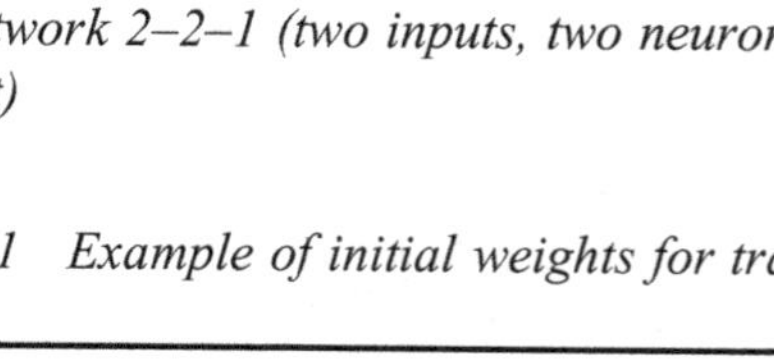

Figure 6.7 Neural network 2–2–1 (two inputs, two neurons in the hidden layer, one output)

Table 6.1 Example of initial weights for training

$w_{11} = 0.1$	
$w_{21} = -0.2$	$w_{01} = 0.2$
$w_{12} = 0.4$	$w_{02} = -0.5$
$w_{22} = 0.2$	

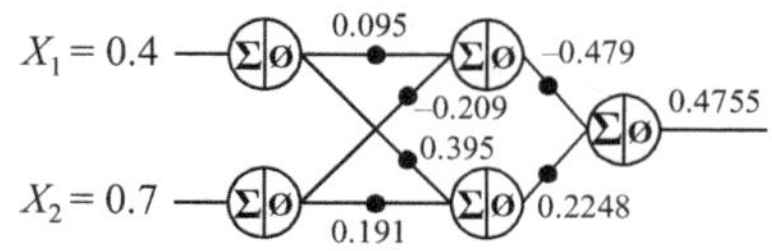

Figure 6.8 Neural network after one cycle of backpropagation

For the weight changes in the hidden layer it is necessary to first calculate $\sum_{k=1}^{n} \omega_{jk} E_k^{output}$

$$\sum \omega_{jk} E_k^{output} = (-0.5)(0.25) + (0.2)(0.25) = -0.075$$

Therefore,

$$\Delta\omega_{11} = (0.7)(0.4)(0.475)(1 - 0.475)(-0.075) = -0.0052$$

$$\omega_{11\ NEW} = 0.1 - 0.0052 = 0.095$$

$$\Delta\omega_{12} = (0.7)(0.4)(0.574)(1 - 0.574)(-0.075) = -0.0051$$

$$\omega_{12\ NEW} = 0.4 - 0.0052 = 0.395$$

$$\Delta\omega_{21} = (0.7)(0.7)(0.475)(1 - 0.475)(-0.075) = -0.009$$

$$\omega_{21\ NEW} = -0.2 - 0.009 = -0.209$$

$$\Delta\omega_{22} = (0.7)(0.7)(0.574)(1 - 0.574)(-0.075) = -0.0089$$

$$\omega_{22\ NEW} = 0.2 - 0.0089 = 0.191$$

And the new configuration for the weights is given in Figure 6.8. With this single pass, the output has come closer to the target value (0.45). A little more training steps and the output will eventually converge to a prescribed error threshold. The previous training example can be further enhanced with the utilization of a bias node and a momentum term.

6.3.2 Error measurement and chain-rule for backpropagation training

An error measure can be defined to quantify the performance of an NN. This error function depends on the weight values and the mapping samples. Determining the weights of an NN can be interpreted as an optimization problem, where the performance error of the network structure is minimized for a representative sample of the mappings. Therefore, it is possible to try randomly generated weight sets and to keep trying with new randomly generated weight sets until one hits it just right. All paradigms applicable to general optimization problems apply therefore to neural nets as well. The random search is at least in principle a way to determine a suitable weight set, but it may require excessive demands on computing time. Smart random search paradigms (such as genetic algorithm and simulated annealing) were used, which legitimate mechanism for NNs. With the availability of ever faster computers allowed, this method can be used. The backpropagation algorithm cannot be applied to every optimization problem, but it can be tailored to multilayer feedforward NN applications. It works really well with one-hidden layer, and often with two hidden layers, as long as the size of dataset does not have a massive number of features, which will require deep learning techniques (discussed in Chapter 9 of this book).

A very common error measure is the MSE. Therefore, E is determined by calculating the output value for each sample, and then computing the differences between the desired target minus the calculated output, in accordance with (6.16), which is based on the square of the Euclidean distance of target to output (although other algebraic norms can be used for distance):

$$E = \frac{1}{2}\sum_{i=1}^{n_{outputs}} (t_i - o_i)^2 = \frac{1}{2}||t_i - o_i||^2 \tag{6.16}$$

Minimization of error means improving performance on the training set and getting a more accurate model. Different definitions and types of error may be considered during training an ANN. For instance, an absolute error is defined as the absolute difference between the measured (actual) output and the desired output (target). However, it is more common to use MSE or root-mean-square error (RMSE) when training MLPs, defined according to (6.17) and (6.18), respectively, where y_m is the measured data, y is the prediction of the model, and n_d is the number of data sets. Minimization of the error can be achieved by changing weights and/or training algorithms:

$$MSE = \frac{1}{n_d}\cdot\sum_{i=1}^{n_d}\left(\frac{y_{mi} - y_i}{y_{mi}}\right)^2 \tag{6.17}$$

$$RMSE = \sqrt{\frac{1}{n_d}\cdot\sum_{i=1}^{n_d}\left(\frac{y_{mi} - y_i}{y_{mi}}\right)^2} \tag{6.18}$$

Training an NN starts out with a randomly selected weight set. A batch of samples is shown to the network, and an improved weight set is obtained by iteration following (6.19) and (6.20). The new weights for a particular neuron (labeled *ij*) at iteration (n+1) are an improvement for the weights from iteration (n), by moving a small amount on the gradient of the error surface toward the direction of the minimum, where η is the learning gain:

$$w_{ij}^{(n+1)} = w_{ij}^{(n)} + \Delta w_{ij} \tag{6.19}$$

$$\Delta w_{ij} = -\eta\frac{\partial E}{\partial w_{ij}} \tag{6.20}$$

Equations (6.19) and (6.20) represent an iterative steepest descent algorithm, which will always converge to a local minimum of the error function, provided that the learning parameter, η is small. The beauty of the backpropagation algorithm is to be a simple analytical expression for the gradient of the error in multilayered nets, calculated by the simple application of the chain rule.

From 1985, for a couple of years it was believed that the backpropagation algorithm was the only practical algorithm to implement such a derivative of the error (E) with respect to the weights, but other training possibilities were developed. The calculation of the derivative of the error in respect to the weights can be easily performed with transfer functions that have derivatives in terms of their own functions, because the derivative can be calculated using a lookup table and making the implementation in real-time very fast.

The transfer (activation) functions are the ones that after taking the weighted sum at the input of the neuron will give an output signal. Two common transfer functions that are employed for MLP are the sigmoid and the tanh (hyperbolic tangent). Section 6.4 and Figure 6.10 have further details. Sigmoid and tanh are differentiable and can cope with the nonlinearity of typical engineering problems. Details for deriving the backpropagation algorithm with the derivative of the transfer functions can be found in the literature.

In order to calculate the partial derivative of the error in respect to the weights for any output neuron, the chain rule can be devised such as (6.21) where $\partial(net\ i)$ is defined as the error term contribution of net-neuron i to the total error:

$$\frac{\partial E}{\partial w_{ij}} = \frac{\partial E}{\partial(net\ i)} \cdot \frac{\partial(net\ i)}{\partial w_{ij}} \tag{6.21}$$

If an output neuron is considered like Figure 6.9 where there are several incoming inputs from the previous neurons, i.e., $y_1 \cdots y_j$, each one multiplied by their corresponding weight $w_{i1} \cdots w_{ij}$, the partial derivative of the error term contribution of the net-neuron neuron i to each weight w_{ij} is given by (6.22), i.e., the weight variation can be conducted proportionally to the incoming input. For a perceptron this is straightforward, but for a multilayer NN, it is necessary to assign an error signal to this input that was not clear for many years. The backpropagation algorithm assigns a "blaming" value, or an error estimation, with a partial derivative of the total error in respect to the error term contribution.

$$\frac{\partial(net\ i)}{\partial w_{ij}} = y_j \tag{6.22}$$

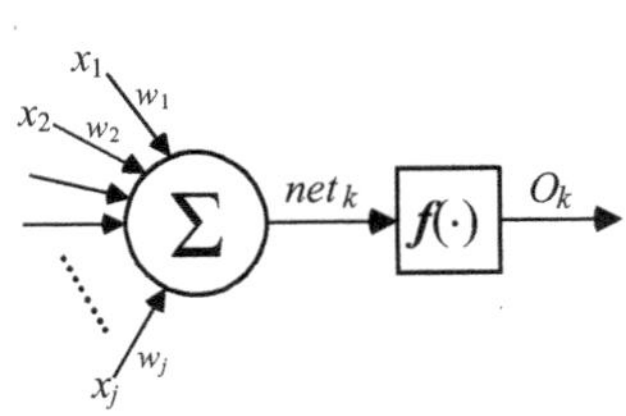

Figure 6.9 A general output neuron with incoming signal x, weighed by w_j, summed and squashed by a transfer function, or activation function $f(\cdot)$, giving the neuron output o_k

Previously we discussed that the error signal term produced by one general output neuron could be expanded using the chain-rule, i.e., calculating the partial derivative of the total error in respect to $(net\ i)$. Therefore, (6.23) defines an estimation for such a kth neuron δ_{ok} error:

$$\delta_{ok} \stackrel{\text{def}}{=} -\frac{\partial E}{\partial(net\ k)} = -\frac{\partial E}{\partial o_k} \cdot \frac{\partial o_k}{\partial(net\ k)} \tag{6.23}$$

The previous chain rule can be expanded as $\partial E/\partial o_k = \left(\partial\left(1/2(t_i - o_i)^2\right)/\partial o_k\right) = -(t_i - o_i)$ and the other term is the derivative of the activation function, i.e., $f_k'(net\ k) \stackrel{\text{def}}{=} (\partial o_k/\partial(net\ k))$; so, (6.24) defines the error signal to be the local error at the output of the kth neuron scaled by a multiplicative factor. This factor is the derivative of the transfer or activation function, i.e., the slope of the function, which should be a smooth and differentiable function in order to make this error estimation possible:

$$\delta_{ok} = (t_i - o_i) \cdot f_k'(net\ k) \text{ for } k = 1, 2, \ldots, k \tag{6.24}$$

Therefore, the complete weight update equation, for a general weight, w_{kj}, in the NN output is given by the following equation:

$$\begin{aligned} \Delta w_{kj} &= -\eta \frac{\partial E}{\partial w_{kj}} = -\eta \frac{\partial E}{\partial(net\ k)} \frac{\partial(net\ k)}{\partial w_{kj}} \\ &= -\eta \frac{\partial E}{\partial(ok)} \frac{\partial(ok)}{\partial(net\ k)} \frac{\partial(net\ k)}{\partial w_{kj}} = \eta(t_i - o_i) f_k'(net\ k) y_j \end{aligned} \tag{6.25}$$

The derivative of the transfer function can be precomputed, and it is usually a very nice feature when such a derivative depends on the own function, so it is easier to implement numerically or in real time. For example, for the *unipolar sigmoidal function*, the activation function is defined by the following equation:

$$f(net) \stackrel{\Delta}{=} \frac{1}{1 + e^{(-\lambda(net))}} \tag{6.26}$$

which is called unipolar sigmoidal transfer function.

Suppose $\lambda = 1$, the derivative $f_k'(net\ k)$ is calculated in (6.27). Depending on the activation or transfer function used in the NN, this derivative must be previously calculated in order to use in the network training, for other types of functions the reader should refer to the literature in this topic.

$$f_k'(net) = \frac{1}{1 + e^{(-net)}} \frac{1 + e^{(-net)} - 1}{1 + e^{(-net)}} \Rightarrow f'(net) = f(net)(1 - f(net)) \tag{6.27}$$

The same reasoning is possible to be made for the hidden layer. However, each neuron at the hidden layer should have a "total blaming error" in order to replace the term previously discussed for the output error. There are several possibilities for calculating this term, but the computation given by (6.28), where there is a weighted average error signal multiplied by the weight that connects the *j*th neuron to the *k*th neuron, is considered a valid generalized error signal for a hidden layer neuron *j*. Equation (6.29) is exactly as (6.25), with the difference that the blaming error (6.28) is used as well as the input for that particular hidden layer neuron coming from the first layer. The training equation is the same, but the error is estimated and the input from the previous layer (sometimes the input of the NN for a topology with just one hidden layer) is used:

$$\delta_{hidden\ layer\ neuron\ j} = \sum_{k=1}^{k} \left(\delta_{ok} \cdot w_{kj}\right) \tag{6.28}$$

$$\Delta w_{kj} = \eta\left(\delta_{hidden\ layer\ neuron\ j}\right) f_k'(net\ k)\left(input_j\right) \tag{6.29}$$

There are two different modes of training in ANNs: incremental training and batch training. In incremental training, weights and biases of the network are updated each time an input is presented to the network. In batch training, the weights and biases are only updated after all inputs are presented. Batch training methods are generally more efficient in production frameworks. However, there are some applications where incremental training can be useful, so that the paradigm is available as well. The training process of an ANN involves the tuning of one or more hyperparameters. For example, it is needed to change the number of neurons in the hidden layer in order to attain the best converging network. The number of neurons is an example of a hyperparameter that changes the complexity of the mapping that can be approximated by the ANN. It is desirable to use the simplest possible network structure with the least number of free parameters (weights). The developed model can be utilized to validate new process measurements.

A complete ANN training procedure is usually based on an iterative approximation in which the parameters are successively updated in numerous steps. Such a step can be based on a single data or a group for all available data points. In each step, the desired outcome is compared with the actual one, and using knowledge of the architecture, all parameters are changed slightly such that the error for the presented data points decreases. Several other algorithms can be used for training ANNs. They are always based on gradient or Jacobian-based methods, sometimes hybrid with Hebbian learning, as the following ones:

- Levenberg–Marquardt
- Bayesian regularization
- BFGS quasi-Newton
- Resilient backpropagation
- Scaled conjugate gradient
- Conjugate gradient with Powell/Beale restarts
- Fletcher–Powell conjugate gradient
- Polak–Ribiére conjugate gradient
- One-step secant
- Variable learning rate gradient descent

6.4 Neuron activation transfer functions

There are many kinds of activation functions for neurons; such an activation or transfer function takes the neuron activation (weighted sum of all inputs and the bias) and generates the neuron's output. Figure 6.10 shows some typical neuron activation (transfer) functions.

A linear transfer function produces an output with a constant slope, in which the gain is the slope of the line. Linear transfer functions are not often used in regular feedforward NNs, because neurons should be able to synthesize nonlinear behavior. The advantage is that the derivative is a constant, very easy to calculate. A linear threshold transfer function is one in which the output is a constant multiple of the input over some range, possibly shifting to the left or to the right by a constant. Below such a range the function takes a constant "low value" and above such a range it takes a constant "high value," i.e., a linear threshold function is a linear transfer function, clipped to either low or high, with a center equal to (high+low)/2, and the slope of the linear portion of the curve is the gain. Such a linear threshold transfer function is capable of synthesizing nonlinear behavior.

When a neuron is supposed to have a binary output, a step transfer function, sometimes also called Heaviside function, is limited to two possible values: below threshold the output is always low, and above threshold it is always high, acting just like a digital logic function. Center is the value of the input at which the output jumps from low to high (if the neuron activation is increasing), or from high to low (if the neuron activation is decreasing). This threshold transfer function was used by McCulloch and Pitts in 1943, and it is typically the activation function used in Hopfield NNs. Backpropagation will work with neurons on step transfer functions as they need a numerical implementation to calculate the derivative of such a transfer function (a step has a mathematical Dirac impulsive solution, which is approximated for computer-based implementation).

A sigmoid transfer function, also called an S-shaped, semi-linear, squashing function, or logistic function, is the one in which the output is continuous, monotonic, and with a continuous derivative that can be written in terms of the original function. It asymptotically approaches a low and a high value, and the center is the value of the input at which the output is equal to (high+low)/2. Gain is directly proportional to the derivative of the function at the center point. The S-function is both nonlinear and continuously differentiable, so the backpropagation algorithm works well. For high gain (slope at the center) the sigmoid approaches a step-function, while at low gain (slope at the center) it approximates a linear function.

A Gaussian transfer function (not depicted in Figure 6.10), also called a *Bell curve*, is not monotonic, but the symmetrical halves are monotonic. It is continuous and continuously differentiable, similar to a sigmoid function. The Gaussian transfer function could be understood as two sigmoids, one an S-shape rising from the left to a peak value, and the other S-shape falling toward the right. The center is the value of the input in which the output is equal to high. The gain is proportional to the standard deviation of the Gaussian. This transfer function may work really

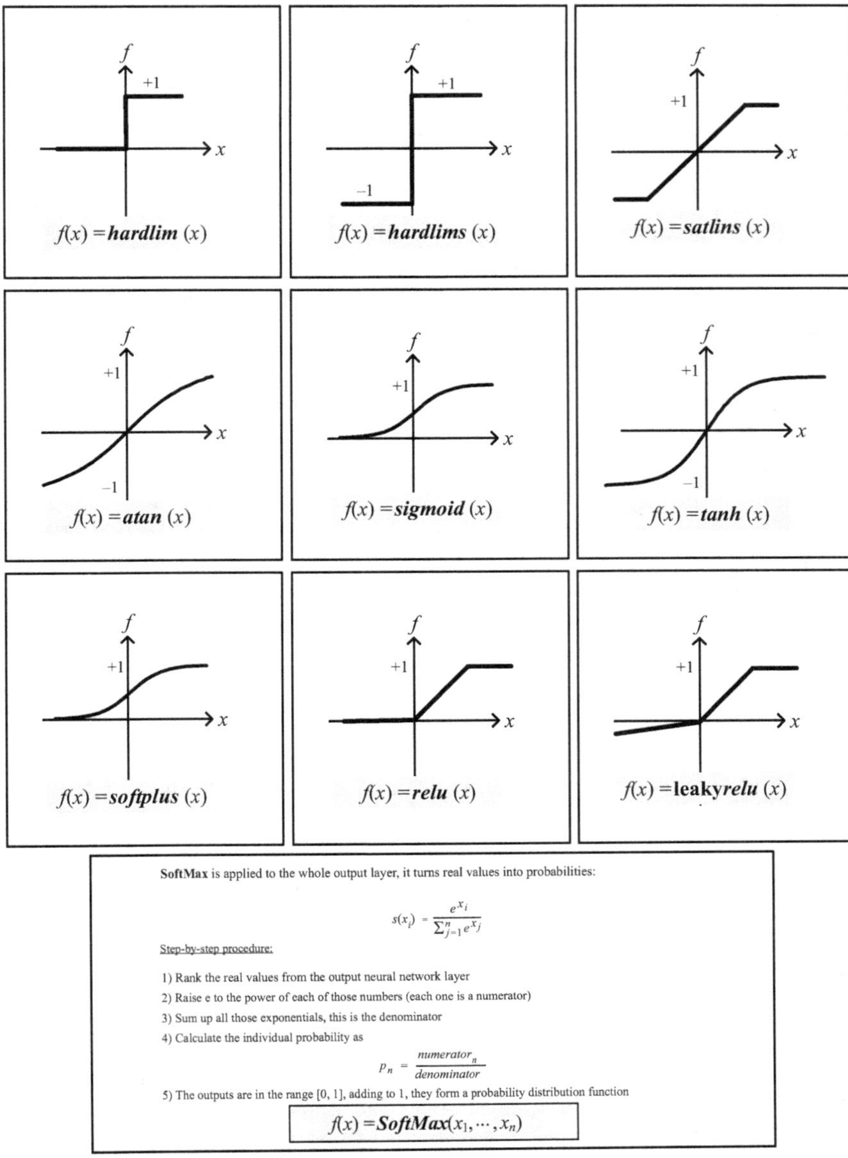

Figure 6.10 Typical neuron activation, transfer functions

well with non-separable data but may not converge with separable data with clear clusters.

A great alternative to sigmoid function is the hyperbolic tangent function, tanh $(x) = (e^x - e^{-x})/(e^x + e^{-x})$. The derivative of a hyperbolic tangent function has a very simple form, which is very useful during backpropagation. This transfer function is symmetrical and bipolar and can be used for continuous numerical calculation. It is very common in NNs and, together with the sigmoidal function, is used in LSTM recurrent NN for deep learning (see Chapter 9). Most of backpropagation feedforward NNs will use either a hyperbolic tangent or the logistic sigmoid function.

During backpropagation in NNs with too many, say N hidden layers, the derivative of the transfer function will multiply the original function, and the error will be squashed backward to the power of N. Since the error is typically way less than 1, it will squash and vanish, making it difficult for NNs with too many hidden layers to achieve convergence during training, typically trying to capture more features of the training data. This problem has been solved by the ReLU activation function, as described in Chapter 9 of this book.

6.5 Data processing for neural networks

A successful NN application requires to have appropriate training data. The training set must span the total range of input patterns sufficiently well, so that the trained network can generalize about the data and not memorize the presented patterns. The first step of data selection is to determine which variables are most used in practice and how they are used. Combining and preprocessing data to make it more meaningful is extremely beneficial. For example, in many applications, the power spectrum density functions are more useful as inputs to NNs than the time series sampled data from time records.

After data are collected, it must be examined to identify any characteristics that may indicate more complex relationships. The calculation of the correlation coefficient for two variables can give an indication of the strength of the relationship between these two variables. If there are two highly correlated input variables then it is possible to reduce the size of the model. An outlier is an extreme data point that may be the result of erroneous data entry. However, outliers are not always errors, which can indeed indicate some sort of anomaly that must be further investigated. A graph of the data is helpful in visualizing trends in data and for identifying outliers. Anomalous outliers can be disruptive in the model development.

How data are represented and translated is important to help the network grasp the problem. Data can be continuous-valued or binary. For example, room temperature could be an input varying from 8°C to 32°C, or it could be represented by five binary inputs: cold, cool, comfortable, warm, and hot. Such inputs could be further improved with fuzzy representations of the previous temperature definitions. Another important issue is whether to use actual input variable values or changes in these input values. In financial forecasting, for example, the amount of money in a banking certificate may be less important than the variation of this amount in a certain period. Another valuable data-processing step that can help the training of the network is the utilization of input variable ratios, especially when it is known first hand that they can influence the output. Although the network can learn to do the division by itself by effectively converting the numerator and denominator of each to logs, it is more productive to carry out the division ahead of time and let the NN concentrate on establishing relationships from the data.

NNs are very sensitive to the absolute scale in the values. If one input is much bigger than other, the network can erroneously assume a higher importance for such a variable. In addition to this magnitude sensitivity, the input data must correspond to the range of the activation function (from 0 to 1, or from -1 to $+1$) to avoid network paralysis in the training process that occurs when weights become very large, thereby forcing the neurons to operate in a region where the activation function is very flat, i.e., its derivative is very small. Since the error sent back for training in backpropagation is proportional to the derivative, very little training would take place. To avoid such problems, numeric input data must be normalized by simply dividing all sample values of the variable by the maximum value, so that the input data are in per-unit basis, and then scaled, so that the minimum and maximum value are within the linear range of the activation function. Figure 6.11 shows a block diagram with the necessary steps to train and develop an NN with software. The utilization of commercial or open-source software packages and libraries can help the development and training of NNs. In some special applications, the designer may want to have control of every detail of the training algorithm. In such cases, it is better to write the entire training code by using a high-level computer language.

For processes that may have a mathematical model, a simulation study and analysis can generate the training data for an NN. Such simulation must be further validated with actual implementation and retrained, if necessary, to attain all the real process features. The initial setup of the network topology is dependent on the modeler's experience. It is best to start with a small number of nodes in the hidden layer and gradually increase the hidden layer size by trial and error. The designer will make several projects of NNs, with different hyperparameters, such as the number of nodes in the hidden layer, and the type of transfer functions. After convergence, the performance of those projects on a test data that was not used for the training set will be compared and considered. After the network has been trained, it is important to test it against the training set and with examples that the network has never met before.

Increasing the size of the hidden layer usually improves the network's accuracy on the training set, but decreasing the size of the hidden layer generally improves generalization, and hence the performance on new cases. It is particularly important to keep the number of layers in the network to a minimum. Every time the error from the output layer is backpropagated to the middle layers, it becomes less and less meaningful, because the correlation of the output layer's errors to the last middle layer according to the connection weights is, in a broad sense, only a guess about what the middle layer's errors actually is.

With an iterative procedure such as the backpropagation algorithm, a question that arises is how and when to stop the iteration. When the performance index (global error) has been reduced to zero or some threshold value, the solution has been ultimately found. However, during the development phase the error will never be small, and two approaches can be used to determine when to stop: (1) limit the

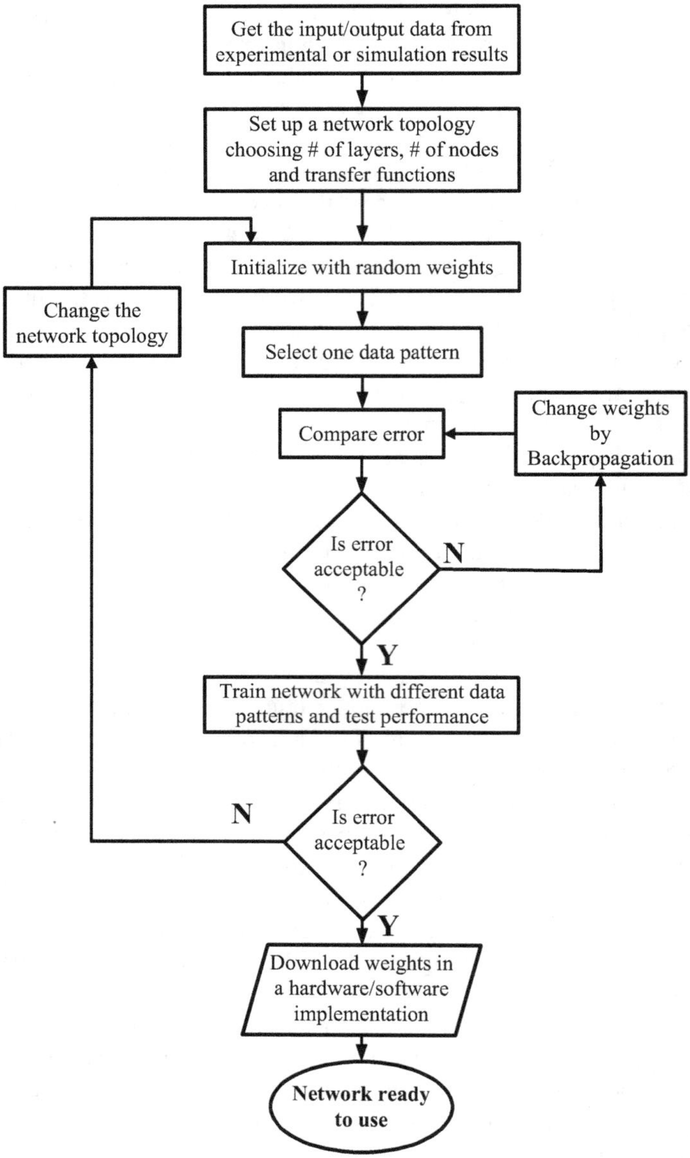

Figure 6.11 Flow chart for neural network development

number of iterations, i.e., the training ceases after a fixed upper limit on the number of training epochs, (2) the error can be sampled and averaged over a fixed interval of epochs, for example, every 500 epochs of training. If the average error for the most recent 500 epochs is not better than that for the previous 500, no progress is being made, and training should halt. After stopping training the network should recall a set of data not actually used during the training phase. If the performance is not satisfactory, the weights may receive a small amount of random noise to help the network get out of the local minimum, or the network can be completely reinitialized.

6.6 Neural-network-based computing

Chapters 3, 4, and 5 in this book discussed the foundations of fuzzy logic (FL), supporting that such a technique is associated with imprecision, approximate reasoning, natural language, and semantics of real-life descriptions as related to modeling and control process, with learning and curve fitting probabilistic reasoning to uncertainty. Zadeh defined FL as "computing with words." The paradigm introduced in this Chapter 6 discusses a computing approach implemented by a family of algorithms defined as ANNs. They were inspired, during the first wave of NN research, by cybernetics and psychology and. On their second wave, these models were inspired by connectionism and parallel computing, as researchers have been trying to understand biological NNs, such as modeling the central nervous systems of animals, particularly the brain, or spine connections, or cerebellar/cortex models. They are used to estimate or approximate functions that can depend on a large number of inputs and are generally unknown. ANNs are generally presented as systems of interconnected "neurons," mathematically represented by connections with numeric weights, which can be tuned based on data, making neural nets adaptive to inputs, and capable of learning. Both paradigms FL and NNs have pros and cons to be used when

- applicable to nonlinear systems;
- have ability to deal with nonlinearity;
- follow more human-like reasoning paths than classical methods;
- utilize self-learning, numerical examples for training;
- FL may integrate yet-to-be-proven theorems with practical rules;
- NNs are robust in the presence of noise, errors, and imperfect data.

This book does not intend to cover all paradigms, topologies, types, and training methodologies for all NNs. The field is very broad, so the reader is encouraged to study more in-depth books, course materials, and scientific papers. However, understanding backpropagation, how to calculate error in order to make training to converge, regular activation functions, and data preparation is more than 90% of times applicable for designing estimators, classifiers, controllers, used in power electronics, power systems, smart-grid, and renewable energy conversion. The next two chapters, i.e., Chapters 7 and 8 will cover on further neural-smith, similar to a blacksmith who has a general knowledge of how to make and repair many things, from the most complex of weapons and armor to simple things like nails or lengths of chain. NNs' design involves math, computer science, engineering modeling and analysis, as well as experience, and a degree of intuition as well.

The next chapters will present a few other topologies and applications in power electronics and power systems. Chapter 9 will concentrate on how deep learning became a new third wave of NN research and its relationship to massive data requirements of smart-grid and modern electrical power systems.

Chapter 7
Feedback, competitive, and associative neural networks

Artificial neural networks (ANNs) is an emerging research area with a wide range of potential applications in science, art, and engineering. It has many advantages over conventional modeling approaches. ANN methodology can be a suitable alternative to classical statistical modeling techniques when obtained datasets indicate nonlinearities in the system. Several combinatorial optimization problems in industrial plants, system identification, and complex uncertain nonlinear systems can be approached by ANN methodology. ANNs can perform nonlinear modeling and filtering data, finding coupled nonlinear relations between independent and dependent variables without their dynamic equations. It is also a cost-effective and reliable approach for condition monitoring, where data related to the condition of the system can be classified and trained; ANNs can be applied to examine condition-based maintenance, detect anomalies, and isolate faults.

The use of ANNs for sensor validation minimizes the need for calibration decreasing the shutdowns due to sensor failure. ANNs have been considered as an acceptable solution to many problems in modeling and control of nonlinear systems. Real data obtained from an industrial system can be used to develop a simple ANN model of the system with very high prediction accuracy. In control design, a neural network (NN) may directly implement the controller (direct design). In this case, the NN will be trained as a controller based on some specified criteria. It is also possible to design a conventional controller for an available ANN model (indirect design). The obtained data from systems located in industrial factories and plants may include noisy data or might be inaccurate or incomplete due to faulty sensors, particularly for aging systems where maintenance is poor. ANNs have the capability to work well even when the datasets are noisy or incomplete. The development of an ANN model requires less formal scientific personnel and does not need professional statistical knowledge. If the datasets and appropriate software are available, even newcomers to the field can handle the NN design and implementation process. They also have the capability of dealing with stochastic variations of the scheduled operating point with increasing data and can be used for online processing and classification.

In addition to the applications of ANNs to industrial systems, they have many general advantages such as simple processing elements, fast processing time, easy training process, and high computational speed. They capture any kind of relation and association, exploring regularities within a set of patterns, and have the capability to be used for a very large number and diversity of data and variables. They provide a high degree of adaptive interconnections between elements and can be used where the relations between different parameters of the system are complex to find with conventional approaches. ANNs are not restricted by assumptions such as linearity, normality, and variable independence, as other conventional techniques. They can generalize in situations for which they have not been previously trained. Generally, it is believed that the ability of ANNs to model different kinds of industrial systems in a variety of applications can reduce the time spent on model development leading to a better performance when compared to conventional techniques.

Chapter 6 covered feedforward ANNs, i.e., structures allowing signals to travel one way only, from input to output. There is no feedback, there are no loops. The output of any layer does not affect that same layer, only the next one. Feedforward ANNs are considered an instantaneous mapping technique, as they associate inputs to outputs. Therefore, those networks are extensively used in pattern recognition. However, in order to train an NN using backpropagation, the output will be used to calculate the changes of the weights at the input of each neuron. The excitation signal only flows from the input towards the output. Figure 7.1 shows a generalized typical feedforward network topology. There are no lateral connections within each layer, and no feedback connections within the network. As explained in Chapter 6, the use of multilayer perceptron (MLP) is a typical implementation. Figure 7.1 shows an ANN with a single hidden layer (it could have two) that is often used for function fitting, pattern recognition, and nonlinear classification. Among different ANN structures, MLP is the first choice for modeling and simulation of nonlinear behavior of industrial systems.

There are other types of feedforward architectures such as the cerebellar model articulation controller (CMAC), radial basis function (RBF) networks, fuzzy NNs (FNNs), adaptive network-based fuzzy inference system (ANFIS), and an adaptive implementation of FNNs. Another powerful topology is the convolutional NN (CNN) that uses several activation functions, with feature-based relationships between weights, CNNs became well adapted for the deep learning era of NN

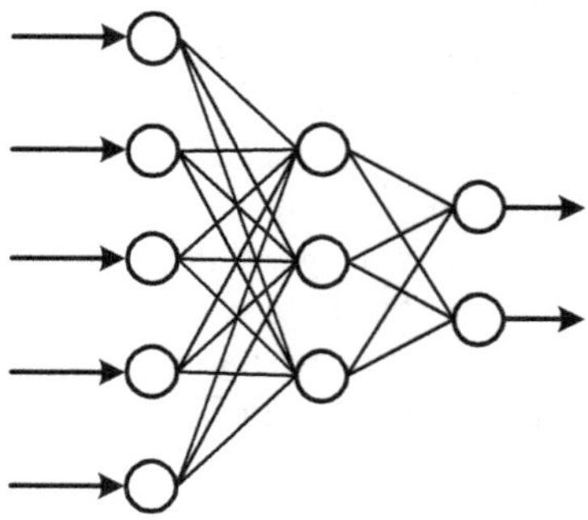

Figure 7.1 Feedforward neural network with three layers, five inputs, and two outputs

research. Feedforward NNs can be implemented in supervised, unsupervised, and reinforcement learning, in single- and multilayer, for the majority of physical and industrial problems. The reader is encouraged to study further in other compendiums and textbooks (Reed, 1999).

7.1 Feedback networks

Feedback networks have signals traveling in both directions by introducing loops in the network structure. Feedback networks are very powerful and may be extremely complicated. Feedback networks are dynamic, i.e., their state may change continuously until reaching an equilibrium point. They remain at the equilibrium point until the input changes, and a new equilibrium needs to be found. A general topology is illustrated in Figure 7.2.

Feedback architectures as indicated in Figure 7.2, are referred as interactive or recurrent. A recurrent network is usually a feedback connection with single-layer organizations; the self-organizing networks (or Kohonen networks) that were rebranded as self-organizing maps (SOMs) are solid implementations of feedback networks. A feedback NN is a dynamic structure that allows modeling of time-domain behaviors of a dynamic system. The outputs of a dynamic system depend not only on present inputs, but also on the history of the system states and inputs, and a recurrent structure is needed to model such behavior. Recurrent networks are a class by themselves, and they are discussed in Chapter 9.

After the publication of the book by Minsky and Papert (1969) describing the limitation of perceptron's training in 1969, the funds for NN research dried out. However, a few different structures were developed, such as the CMAC by Albus (1975), and later the principles and foundation for the SOMs and associative memory by Kohonen in 1982, and the Hopfield network (Hopfield, 1982; Hopfield and Tank, 1986) in 1982 and 1986. There were independent development of backpropagation by three different groups and individuals, Amari (1967), Werbos (1974), and Rumelhart *et al.* (1986) that made possible a rebirth of the ANN research around 1985/1986. Particularly, relevant to feedback–competitive–associative NNs, the seminal paper by Teuvo Kohonen, titled "Self-Organized Formation of Topologically Correct Feature Maps" (Kohonen, 1982), showed an algorithm that would effectively map similar patterns (pattern vectors close to each other in the input signal space) onto contiguous locations in the output space,

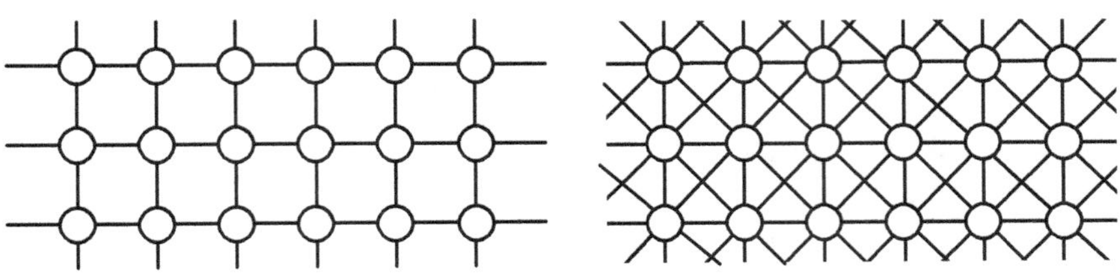

Figure 7.2 Feedback, competitive, and associative neural networks

expanding previous algebraic mathematical correlation principles (Kohonen, 1972, 1974) on similarity (clustering) diagrams for the map of phonemes and speech recognition in 1983. The initial Kohonen Maps were developed on supervised learning, and a few years later they became established as SOMs (Kohonen *et al.*, 1990).

A methodology to train any NN, feedforward or feedback, is portrayed in Figure 7.3. Initially it is necessary to collect or generate the data to be used for training and testing the NN. Once data is gathered, it must be divided into a training and a test set. The training set should cover all the input space or at least cover the specific region in which the network is expected to operate. If there are no training data for certain conditions, the output of the network should not be trusted for such conditions. The split of data into training and test sets is dependent on the knowledge of the system, and sometimes it is based on a trial-and-error procedure. In one hand, the training set should be small in order to have a fast training, on the other hand, the input space must be well covered, which may require a large training set.

Once the training set is selected, the NN architecture must be chosen. There are two possibilities: (i) some designers may start with a fairly large network that is sure to have enough degrees of freedom (neurons in the hidden layer) in order to train to the desired error goal; then, once the network is trained, it will be pruned or shrunken until the smallest configuration maintains the same performance; (ii) other designers may start with a small network and increase it until the network trains, and the error goal is met.

After the network architecture is chosen, the weights and biases are initialized, and the network is trained. The network may not reach the error goal due to one or more of the following reasons:

- Training gets stuck in a local minimum.
- The weights and biases are reinitialized, and training is restarted.

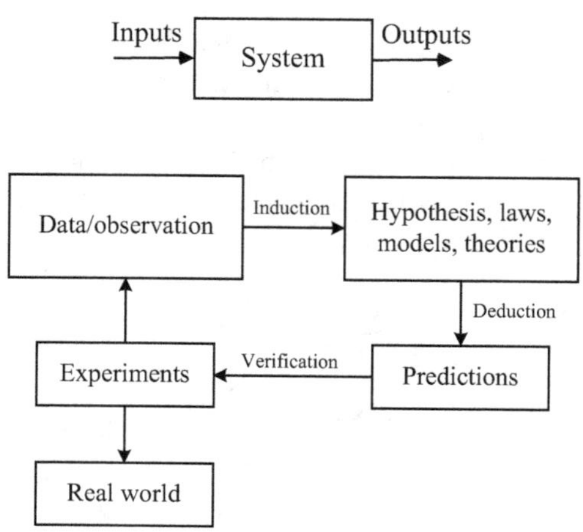

Figure 7.3 Neural network design methodology, modeling based on data validation

- The network may not have enough degrees of freedom to fit the desired input/output model, and probably more neurons must be added to the hidden layer.
- Additional hidden nodes or layers might be added, and network training is restarted.

- There is not enough information in the training dataset to perform the desired mapping.
  - When attempting to train an NN, the architecture that trains correctly (meets the error goal) must respond well to the test set, otherwise it may have overfit. Overfitting is not a good outcome and must be avoided. Therefore, a balance must be met in achieving the minimum error for training as well as for testing.

The cause of the poor test performance is evaluated by using cross-validation checking. If an incomplete test set is causing the poor performance, the test patterns that have high error levels should be added to the training set, a new test set should be chosen, and the network should be retrained. If there are not enough data left for training and testing, it may need to be collected again or regenerated. These training decisions will be covered in more detail and augmented with examples. NN training data should be selected to cover the entire region of the input space where the network is expected to operate. Usually, a large amount of data are collected, and a subset of that data is used to train the network. Another subset of that data is then used as test data to verify the correct generalization of the network. If the network does not generalize well on several data points, that data subset is added to the training dataset, and the network is retrained. This process continues until the performance of the network is acceptable.

Chapter 6 discussed the principles of feedforward NNs, assuming that most of time they are used for mapping inputs/outputs. In the second wave of NN research, connectionism allowed researchers to develop hundreds of different NN topologies and training algorithms. A good discussion on topologies, learning, and functionalities for practical use until the middle of the 2000s has been given in the study of Meireles *et al.* (2003). In addition to feedforward MLPs, the Hopfield network and the Kohonen/SOM network were also important paradigms introduced in the 1980s which became mature for industrial applicability.

The physicist Hopfield (1982) wrote a paper about how NNs form an ideal framework to simulate and explain the statistical mechanics of phase transitions. The Hopfield network can also be viewed as a recurrent content addressable memory that can be applied to image recognition and traveling-salesman-type of optimization problems. For several specialized applications, this type of network is far superior to any other NN approach. On another successful research, Kohonen from Finland proposed an NN, known by his name, which is a one-layer feedforward network that can be viewed as a self-learning implementation of the K-means clustering algorithm for vector quantization (VQ) with powerful self-organizing properties and biological relevance.

There are other powerful and interesting NN paradigms such as the RBF network, the Boltzmann machine, the counterpropagation network (CPN), and the

ART (adaptive resonance theory) networks. RBFs can be viewed as a general regression technique for multidimensional function approximation which employ Gaussian transfer functions with different standard deviations. The Boltzmann machine is a recursive, simulated annealing type of network with arbitrary network configuration. Hecht-Nielsen's CPN combines a feedforward NN structure with a Kohonen layer. Grossberg's ART networks use a similar idea but can easily be implemented in hardware. There are also specialized networks such as Oja's rules for principal component analysis, wavelet networks, cellular automata networks, and Fukushima's neocognitron (Fukushima *et al.*, 1983). Wavelet networks utilize the powerful wavelet transform and usually combine elements of the Kohonen layer with RBF techniques. Cellular automata networks are an NN implementation of the cellular automata paradigm, popularized by Stephen Wolfram (developer of the Mathematica software package). Fukushima's neocognitron is a multilayered network with weight sharing and feature extraction properties that achieved the best performance for handwriting and OCR recognition applications at that time.

A variety of higher order methods improve the speed of the backpropagation algorithm. Most widely applied are conjugate gradient networks and the Levenberg–Marquardt algorithm. Recursive networks with feedback connections are being increasingly adopted, especially in neuro-control problems. For control applications, specialized and powerful NN paradigms have been developed, and it is worthwhile noting that a one-to-one equivalence can be derived between feedforward neural nets of the backpropagation type and Kalman filters. Fuzzy logic and NNs are often combined for control problems. There is no shortage of NN tools, and most paradigms can be applied to a wide range of problems. Most NN implementations rely on the backpropagation algorithm, but there are enhancements and other possibilities.

7.2 Linear Vector Quantization network

Linear Vector Quantization (LVQ) was suggested by Kohonen as an extension of his original work in SOMs. Both are based on a competitive layer, called Kohonen layer, that has the computational capability of sorting items, associating similar objects with their categorized classes. Linear Vector Quantization is an ANN model used not only in multiclass classification problems, but also in many other applications such as image segmentation problems and identification of acoustic wave data transmission over pipelines (Simões *et al.*, 1998, 2000).

The LVQ network contains an input layer and a single Kohonen (or competitive) layer, as portrayed in Figure 7.4. The Kohonen layer has a number of processing elements for each of these classes. The number of processing elements per class depends upon the complexity of the input–output relationship. Usually, each class will have the same number of elements throughout the layer. The Kohonen layer learns and performs correlations and classifications on the training set, using supervised learning, adapted from backpropagation. In order to enhance learning and recall functions, the input layer must contain one processing element for each separable input parameter.

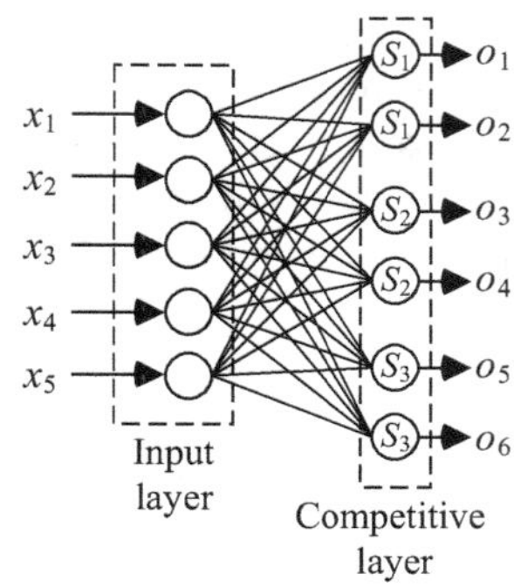

Figure 7.4 Linear Vector Quantization network example topology for a 3-class classification problem (classes S_1, S_2, and S_3). In this particular example, the input layer has five units, so the input vectors should be five-dimensional. The competitive (or Kohonen) layer has six neurons, being two units for each class. The weight vectors connecting the input to each Kohonen unit are called codevectors, and the group of codevectors of the same class is called the codebook of this class

Kohonen networks or SOMs are a competitive training NN aimed at ordering the mapping of the input space. In competitive learning, there is a distributed input vector $x(t)$ sampled at each time t, and a set of reference vectors m_c, usually randomly initialized. The algorithm will minimize a loss function by finding the set of reference vectors that best fit the input set. For each input vector, the network calculates the Euclidean distance between this vector and all reference vectors m_c, declaring the closest one as the "winner" m_{c^*}, which will eventually represent the form of the input as much as possible. If the input vector has a given probability density function $p(x)$, the training process will minimize the average Euclidean distance at the r-power between the input space and the set of reference vectors m_c, as represented in (7.1). This methodology is known as vector quantization (VQ), and the distance is called quantization error in this context.

$$E = \int \|x - m_c\|^r p(x)dx \tag{7.1}$$

The networks can be trained to classify inputs while preserving the inherent topology of the training set. Topology preserving maps preserve the nearest neighbor relationships in the training set such that input patterns that have not been previously learned will be categorized by their nearest neighbors in the training data.

During training, the Kohonen layer of this supervised network computes the distance of a training vector $x(t)$ to each processing element m_c, and the nearest processing element is declared the winner, indicated by index c^*. There is only one winner for the whole layer. The winner will be the only output processing element to fire, announcing its class or category S_{k^*} as the estimated class, indicated by the class index k^*. The estimated class index is compared to the target class index $y(t)$.

If the estimated and the target classes are the same, the weight vector of the winner is rewarded by being updated toward the training vector. Otherwise, if the winning element is not in the target class, its connection weights are punished by being moved away from the training vector. All the other connection weights are left intact. This rule is described in (7.2). During this training process, individual processing elements assigned to a particular class migrate to the region associated with their specific class.

$$m_c(t+1) = \begin{cases} m_c(t) + \alpha(t) \cdot d[x(t), m_c(t)], \text{ if } c = c^* \text{ and } y(t) = k^* \\ m_c(t) - \alpha(t) \cdot d[x(t), m_c(t)], \text{ if } c = c^* \text{ and } y(t) \neq k^* \\ m_c(t), \text{ if } c \neq c^* \end{cases} \quad (7.2)$$

During the recall mode, the distance of an input vector to each processing element is computed, and again the nearest element is declared the winner class of which is considered the estimated class.

There are some shortcomings with the learning VQ architecture. Obviously, for complex, classification problems with similar objects or input vectors, the network requires a large Kohonen layer with many processing elements per class. This can be overcome with selectively better choices for, or by using higher order representations of the input parameters. The basic learning mechanism, called LVQ1, has some weaknesses that have been addressed by variants to the paradigm (namely, OLVQ1, LVQ2.1, and LVQ3). Normally these variants differ from the basic algorithm in different phases of the learning process. They imbue a conscience mechanism, a boundary adjustment algorithm, and an attraction function at different points while training the network. The simple form of the learning VQ network suffers from the defect that some processing elements tend to win too often, while others, in effect, do nothing. This particularly happens when the processing elements begin far from the training vectors. Here, some elements are approximated very quickly, while others remain permanently far away. To alleviate this problem, a conscience mechanism is added so that a processing element that wins too often develops a "guilty conscience" and is penalized. The actual conscience mechanism is implemented by a distance bias which is added to each processing element and is proportional to the difference between the win frequency of an element and the average processing element win frequency. As the network progresses along its learning curve, this bias proportionality factor needs to be decreased.

The boundary adjustment algorithm is used to refine a solution once a relatively good solution has been found. This algorithm affects the cases when the winning processing element is in the wrong class and the second best processing element is in the right class. A further limitation is that the training vector must be near the boundary between these two processing elements. The winning wrong processing element is moved away from the training vector, and the second-place element is moved toward the training vector. This procedure refines the boundary between regions where poor classifications commonly occur. In the early training of the learning VQ network, it is sometimes desirable to turn off repulsion. The winning processing element is only moved toward the training vector if the training vector and the winning processing element are in the same class. This option is particularly helpful when a processing element must move across a region having a different class in order to reach the region where it is needed.

The combination of a Kohonen network and a Grossberg outstar, as shown in Figure 7.5, makes a powerful network that can function as an adaptive lookup table in pattern recognition, pattern completion, and signal enhancement. It contains a supervised learning process by virtue of the association of input vectors with the corresponding output vectors. Even though it is as robust as a regular back-propagation NN, it has rapid training and saved computational time, via the construction of a statistical model of the input vector environment.

SOMs allow signal representations to be automatically mapped onto a set of output responses in a way that the responses acquire the same topological relationships as that of the primary events. The self-organized map, an architecture suggested for ANNs, has been used in simulation experiments and practical applications. SOMs have the property of effectively creating spatially organized internal representations of various features of input signals and their abstractions. As a result, the self-organization process can discover semantic relationships in sentences, semantic maps, and brain maps, for example. The SOM algorithm focuses on best matching cell selection and adaptation of the weight vectors. In supervised tasks, the SOM algorithm can be used to initialize the output vectors, which can then be fine-tuned with learning VQ. The use of SOMs in practical speech recognition and semantic mapping has been reported and very successful.

A speaker recognition system based on SOM NNs was presented in Mafra and Simões (2004). The system could achieve more than 99% accuracy in text-independent mode when trained with approximately 17.5 s of voice samples from each speaker and tested with utterances longer than 2.8 s. The voice of each speaker was modeled by an SOM NN, trained to be a specialist in quantizing the feature vectors (VQ) extracted from his voice. The Mel-frequency cepstral coefficient (MFCC) vectors were used as feature vectors, extracted from segments of the voice samples. When a new voice sample was presented, its MFCC vectors were

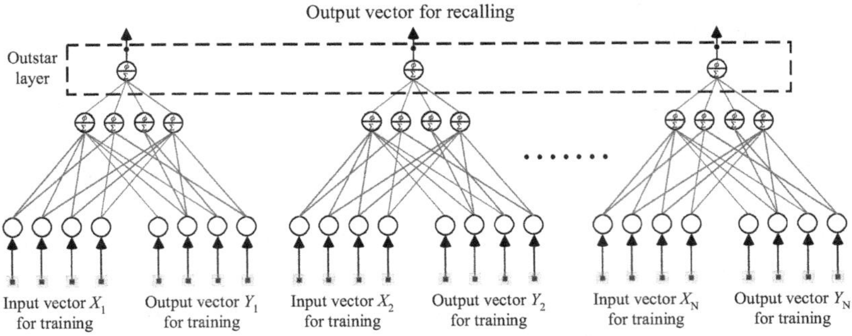

Figure 7.5 Linear Vector Quantization network

extracted and quantized by all SOMs that competed for the speaker: the network that produced the smallest quantization error was declared the winner, defining the recognized speaker. This network ensemble was tried in a speaker identification task, within a closed set of speakers, in text-independent mode. A corpus of voice samples was recorded, taken from 14 speakers (6 men and 8 women), speaking four distinct phrase sets. The first set had variable phrases (answers to personal questions), and the other three sets comprised phonetically balanced phrases in Portuguese. Four SOM architectures were experimented, with 16, 25, 36, and 64 output units. Every combination of a phrase set and an architecture were trained and tested. The results indicated that architectures with more units had more discriminating power, achieving lower quantization errors during training and better precision during tests. It was also seen that longer duration, uniform, and phonetically balanced training sets favored higher correct identification rates. Also, longer samples allowed better identification, whereas the reduction of the sample duration implied a very fast growth of the identification error rates. The proposed architecture showed a highly desirable feature for real-world applications: if a new speaker must be added to the current speaker set, it is only necessary to train a new SOM representing this speaker. There is no need for retraining the already set up networks. This high uncoupling degree makes the method specially interesting when the exact speaker set may vary during the lifecycle of the application. However, an estimate of the maximum number of simultaneous speakers must be provided for the correct dimensioning of the SOMs in order to achieve the desired performance level.

7.3 Counterpropagation network

Hecht-Nielsen (1987) developed the CPN that is a hybrid of the Kohonen and the Grossberg networks, combining unsupervised learning with a supervised output layer. The CPN functions as a statistically optimal self-programming lookup table. It is a different topology but useful for synthesizing complex classification problems. The operation of the CPN is similar to that of the learning VQ network, in that the middle Kohonen layer acts as an adaptive lookup table, finding the closest fit to an input stimulus and outputting its equivalent mapping. The first CPN consisted of a bidirectional mapping between the input and output layers. In essence, while data are presented to the input layer to generate a classification pattern on the output layer, the output layer, in turn, would accept an additional input vector and generate an output classification on the network's input layer. The network got its name from this counterposing flow of information through its structure. Most developers use a uniflow variant of this formal representation of counterpropagation. In other words, there is only one feedforward path from input- to output layer. An example network is shown in Figure 7.6, where the unidirectional CPN has three layers. A fourth layer is sometimes added if the inputs are not already normalized before they enter the network.

The main layers include an input buffer layer, a self-organizing Kohonen layer, and an output layer that uses the delta rule to modify its incoming connection weights. Sometimes this layer is called a Grossberg outstar layer (Simões *et al.*, 2000). For this NN topology, it is important for the designer to have an idea of how many separable parameters would define the problem, because the dimension of the input layer depends on such sizing. It is maybe required a trial-and-error approach, or probably some statistical clustering, in case the designer does not know the physics, or engineering, or science that would support the definition of the separable parameters. For example, a distribution feeder in urban community may have the load and generation cycles very well defined, so the possible faults are known a priori for an NN classification. On the other hand, an underwater autonomous vehicle searching for objects 2,000 m in deep ocean may not have such a priori knowledge. Deciding on the size of the input layer is very critical: if the nodes are too few, the network would not achieve good generalization; if otherwise the nodes are too many, the training process may take too long or even not converge.

For these VQ types of NNs, the input vector must be normalized and fit in accordance to the weight vector scales. Therefore, the Euclidean norm of the input vector must be normalized to the unit. A preprocessor can be used at the input of a CPN or LVQ network, so the input data are properly normalized. This is the same as assuming that a normalization layer could be added between the input and the Kohonen layers. The normalization layer requires one processing element for each input, and an extra one to act as a balancing element. This layer modifies the input vectors before reaching the Kohonen layer to guarantee that all input vectors have unit norm. Without normalization, larger input vectors bias the Kohonen processing elements in a way that weaker input vectors will not be properly classified. The reader can imagine an x–y plane with a vector at 45° and magnitude 100, and another vector at −90° with magnitude 1. This pair cannot be properly identified as two

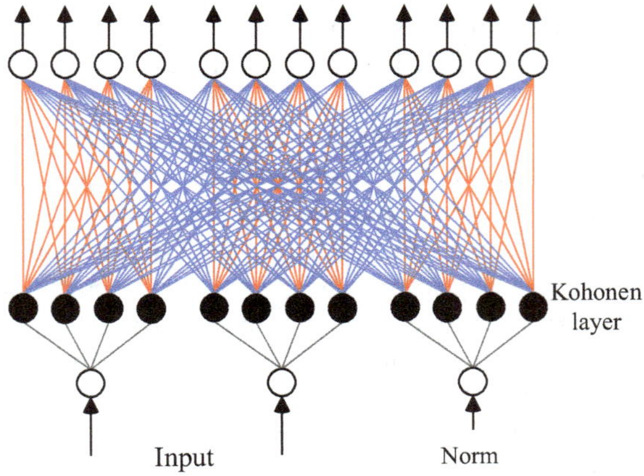

Figure 7.6 Counterpropagation neural network

possible main direction representatives. However, if they are properly normalized to the unit, those directions are clear and may indicate features or classes of the problem.

The CPN uses the standard Kohonen paradigm that self-organizes the input sets into classification zones by following the classical Kohonen learning law. The Kohonen layer acts as a nearest neighbor classifier in that the processing elements in the competitive layer autonomously adjust their connection weights to divide up the input vector space to adapt to the frequency in which the inputs occur. There needs to be at least as many processing elements in the Kohonen layer as output classes. The Kohonen layer usually has a lot more elements than classes because additional processing elements provide a finer resolution between similar objects.

The Grossberg output layer is made up of processing elements that learn to produce a specific output when a particular input is applied. Since the Kohonen layer is competitive, only a single output is activated for each input vector, providing a way to encode that input into a meaningful output class. The Grossberg layer then uses the delta rule to adjust the connection weights of the output layer, while the Kohonen layer is frozen. Since only one neuron from the competitive Kohonen layer is active at a time, while all other elements' outputs are zero, the only connection weights adjusted are the winner's weights in the competitive layer. In this way, the output layer learns to reproduce a certain pattern for each active processing element in the competitive layer. The competitive units belonging to the same class of the winner are also not updated.

The competitive Kohonen layer learns without supervision, it converges without knowing which class it will be associated with in the final response. Therefore, a processing element in the Kohonen layer may be associated with training inputs that belong to different classes. Any input that activates this processing element will produce an ambiguous output. To alleviate this problem, the processing elements in the Kohonen layer could be preconditioned to learn only about a particular class.

A CPN model was developed for the estimation of acoustic signal transmission in an oil-pipeline extraction system (Simões *et al.*, 2000), where the well pressure and temperature were sensed and sent to the top of the line by the ultrasonic signals that performed an frequency shift keying (FSK) type of digital modulation, with two frequencies within the ultrasonic transducer range. The oil pipeline has junctures around every 10 m which cause multireflection and nonlinear distortion, resulting in a complex acoustic signal in the time- and frequency-domains. The usual correlation techniques are not efficient enough for demodulation, and one needs several field measurements for setting up the channel communication model. The counterpropagation NN was used to classify the acoustic signals into two classes of binary transmission, overcoming such nonlinearities and model settings. A preprocessing method of raw input acoustic signal was proposed with a chaotic discrete mapping, so as to build a three-dimensional image of the incoming signal, relating the surface to the time signal. The system was tested in a laboratory with a 100-m long oil pipeline. Acoustic signals were generated by a 150-W piezoelectric transducer, and an accelerometer was responsible for signal reception and delivery to an analog–digital converter interfaced to a PC. The NN algorithm was written in

C and ran in a PC for training and estimation. The input data signals were deliberately corrupted with 40% of noise, and the network was still able to correctly estimate original data with 97.5% accuracy, demonstrating its robustness and stability. The system opened new horizons for oil well monitoring systems. Periodically, the monitoring system would send known patterns for the equipment on the top of the pipeline, which would recalibrate the CPN-based estimation algorithm, making the system parameter insensitive and more reliable in regard to the natural environmental degradation. It is important to observe that in an acoustic transmission system with an NN capability like this, the channel communication model is not necessary. The CPN-based system could be easily adapted to any other oil well, and even retrofitted to the existing ones. The system was fitted into a Brazilian sea oil extraction system.

7.4 Probabilistic neural network

The probabilistic NN was developed by Specht (1988); this network provides a general solution to pattern classification problems by following the approach defined as Bayesian classifiers. Toward the end of the 1990s and the first few years in the 2000s, there were some spin-offs on this ideas, called "Bayesian networks" (Riascos *et al.*, 2007, 2008). Bayes' theory was developed in the 1950s, considering the relative likelihood of events and using a priori information to improve prediction. The network paradigm also uses Parzen Estimators that were developed to construct the probability density functions required by Bayes' theory.

The probabilistic NN uses a supervised training set to develop distribution functions within a pattern layer. These functions, in the recall mode, are used to estimate the likelihood of an input feature vector being part of a learned category, or class. The learned patterns can also be combined or weighted with the a priori probability, also called the relative frequency, of each category to determine the most likely class for a given input vector. If the relative frequency of the categories is unknown, then all categories can be assumed to be equally likely, and the determination of category is solely based on the closeness of the input feature vector to the distribution function of a class.

An example of a probabilistic NN is shown in Figure 7.7, a network with three layers, which contains an input layer that has as many elements as there are separable parameters needed to describe the objects to be classified. It has a pattern layer that organizes the training set such that each input vector is represented by an individual processing element. And finally, the network contains an output layer, called the summation layer, that has as many processing elements as there are classes to be recognized. Each element in this layer combines via processing elements within the pattern layer which relates to the same class and prepares that category for output. Sometimes a fourth layer is added to normalize the input vector, if the inputs are not already normalized before they enter the network. As with the CPN, the input vector must be normalized to provide proper object separation in the pattern layer.

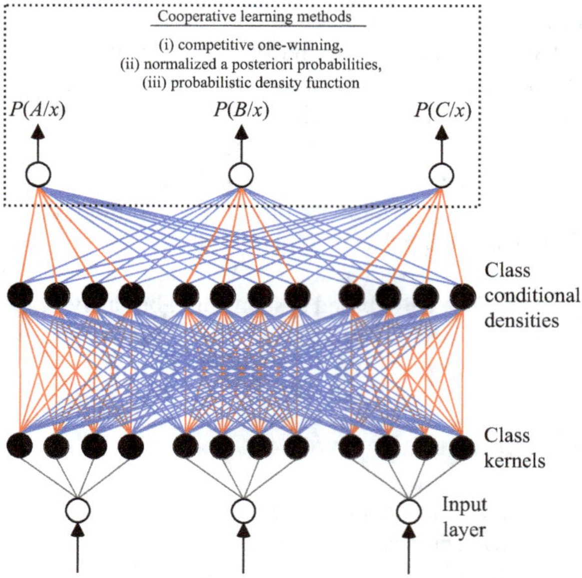

Figure 7.7 Probabilistic neural network

The pattern layer represents a neural implementation of a version of a Bayes classifier, where the class-dependent probability density functions are approximated using a Parzen estimator. This approach provides an optimum pattern classifier in terms of minimizing the expected risk of wrongly classifying an object. With the estimator, the approach gets closer to the true underlying class density functions as the number of training samples increases, so long as the training set is an adequate representation of the class distinctions. In the pattern layer, there is a processing element for each input vector in the training set. Normally, there are equal amounts of processing elements for each output class. Otherwise, one or more classes may be skewed incorrectly, and the network will generate poor results. Each processing element in the pattern layer is trained once. An element is trained to generate a high output value when an input vector matches the training vector. The training function may include a global smoothing factor to better generalize classification results. In any case, the training vectors do not have to be in any special order in the training set, since the category of a particular vector is specified by the desired output of the input. The learning function simply selects the first untrained processing element in the correct output class and modifies its weights to match the training vector.

The pattern layer operates competitively, where only the highest match to an input vector wins and generates an output. In this way, only one classification category is generated for any given input vector. If the input does not relate well to any patterns programmed into the pattern layer, no output is generated. The Parzen estimation can be added to the pattern layer to fine-tune the classification of objects by adding the frequency of occurrence of each training pattern and its corresponding processing elements. Basically, the probability distribution of occurrence for each example in a class is multiplied in its respective training node. In this way, a more accurate expectation of an object is added to the features, which makes it recognizable as a class member.

Training of the probabilistic NN is much simpler than with backpropagation. However, the pattern layer can be quite huge if the distinction between categories is varied, and at the same time quite similar are special areas. There are many proponents for this type of network, since the groundwork for optimization is founded in well-known, classical mathematics.

Chapter 9 has further details on typical machine learning systems and how deep learning can be associated with (i) classification, (ii) regressions, and (iii) clustering. During years 2006–08, three papers were published about Bayesian NNs (Riascos *et al.*, 2006, 2007, 2008) proposing modifications to the original paper called "Probabilistic neural network for fuel cells fault analysis and management control." A Bayesian network is a structure that graphically models relationships of probabilistic dependence within a group of variables. The network structure and the conditional probabilities compose a Bayesian network. A directed acyclic graph represents the structure, where each node is associated to an input variable. Each node has a set of parents, and the conditional probabilities numerically capture the probabilistic dependence among the experimental observed signals. The construction of a graph to describe a diagnostic process can be conducted in two ways: (i) based on human knowledge about the process, where relationships among variables are established to define the criteria for choosing the next state (i.e., the relationship between variables and parents); or (ii) based on probabilistic methods using databases of records. Although a lot of promising results were obtained in Riascos *et al.* (2006, 2007, 2008), it was observed that in the following years there was a decline of interest in *regular* NNs, marking the waning of the second NN research wave, and the emergence of deep learning and big data analytics, or the third NN research wave.

7.5 Industrial applicability of artificial neural networks

Practical considerations regarding the network accuracy, robustness, and implementation issues must be addressed for real-world implementation (Meireles *et al.*, 2003). It has been a constant challenge for researchers to find optimal ANN-based solutions to design, manufacture, develop, and operate new generations of industrial systems as efficiently, reliably, and durably as possible. Getting enough information about the system that is to be modeled is the first step in system identification and the modeling process. Besides, a clear statement of the modeling objectives is necessary for making an efficient model. Industrial systems may be modeled for condition monitoring, fault detection and diagnosis, sensor validation, system identification or design, and optimization of control systems. ANNs have the power to solve many complex problems. They can be used for function fitting, approximation, pattern recognition, clustering, image matching, classification, feature extraction, noise reduction, extrapolation (based on historical data), and dynamic modeling and prediction. ANN-based model building process includes system analysis, data acquisition and preparation, network architecture, as well as network training and validation. For a deeper understanding of the different challenges in using ANN-based methodologies for industrial systems and their applications, advantages, and limitations, the reader is encouraged to read further literature in books and articles.

Fuzzy-logic systems and NNs share the following features:

- estimate functions from sample data;
- do not require mathematical model;
- are dynamic systems;
- can be expressed as a graph that is made up of nodes, and edges convert numerical inputs to outputs;
- process inexact information inexactly;
- have the same state space;
- produce bounded signals;
- a set of n neurons defines n-dimensional fuzzy sets;
- learn some unknown probability function $p(x)$;
- can act as associative memories; and
- NNs can model any system provided the number of nodes, and hidden layers are sufficient.

For ANN applications, it is usually considered a good estimation performance when pattern recognition achieves more than 95% of accuracy in overall and comprehensive data recalls (Riascos *et al.*, 2008). Selection and configuration of the network parameters demand special attention, since it is desirable to use the simplest network while maintaining an adequate performance. Pruning algorithms try to make ANNs simpler by trimming unnecessary links or units, so the cost of runtime, memory, and hardware implementation can be reduced, and generalization power improved. Depending on the application, some functional characteristics of the system are important in the choice of the ANN-topology (Meireles *et al.*, 2003). Table 7.1 summarizes the most common ANN structures used for pattern recognition, associative memory, optimization, function approximation, modeling and control, image processing, and classification purposes. While there are several tutorials and reviews discussing the full range of ANN topologies, learning methods, and algorithms, the following section intends to cover what has actually been applied to industrial applications. It is important to capture such an initial historical perspective of the age of industrial applications, triggered by DARPA Neural Network Study (U.S.) (1988).

Meireles *et al.* (2003) give a historical perspective of both the first and the second waves of research in NNs and successful industrial applications. The first wave started around the Second World War, in 1943, when McCulloch and Pitts proposed the idea that a mind-like machine could be manufactured by interconnecting models based on behavior of biological neurons, laying out the concept of neurological networks. Wiener gave this new field the popular name cybernetics,

Table 7.1 ANN structures used for pattern recognition, associative memory, optimization, function approximation, modeling and control, image processing, and classification purposes

Functional characteristics	Structure
Pattern recognition	MLP, Hopfield, Kohonen, PNN
Associative memory	Hopfield, Recurrent MLP, Kohonen
Optimization	Hopfield, ART, CNN
Function approximation	MLP, CMAC, RBF
Modeling and control	MLP, Recurrent MLP, CMAC, FLN, FPN
Image processing	CNN, Hopfield
Classification and clustering	MLP, Kohonen, RBF, ART, PNN

principle of which is the interdisciplinary relationship among engineering, biology, control systems, and brain functions. Computer architecture was not fully defined by that time, and the research led to what is today known as the von-Neumann-type computer. With the progress in research on the brain and computers, the objective changed from the mind-like machine to manufacturing a learning machine, for which Hebb's learning model was initially proposed.

In 1958, Frank Rosenblatt from the Cornell Aeronautical Laboratory put together a learning machine called the "perceptron" that was the predecessor of most of the NNs. He gave specific design guidelines used by the early 1960s. Widrow and Hoff proposed the "ADALINE" (ADAptive LINear Element), a variation on the perceptron based on a supervised learning rule (the "error correction rule" or delta rule) that could learn in a faster and more accurate way; synaptic strengths were changed in proportion to the error (difference between the actual output and what it should have been) multiplied by the input. Such a scheme was successfully used for echo cancellation in telephone lines and is considered to be the first industrial application of ANNs. During the 1960s, the forerunner for current associative memory systems was the work of Steinbuch (Steinbuch and Piske, 1963) with his "Learning Matrix" that was a binary matrix accepting a binary vector as input, producing a binary vector as output, and capable of forming associations between pairs with a Boolean Hebbian learning procedure.

The perceptron generated considerable excitement when it was first introduced because of its conceptual simplicity. The ADALINE is a weighted sum of the inputs together with a least-mean-square (LMS) algorithm to adjust the weights and to minimize the difference between the desired signal and the actual output. Because of the rigorous mathematical foundation of the LMS algorithm, the ADALINE became then a powerful tool for adaptive signal processing and adaptive control, leading to work on competitive learning and self-organization. Minsky and Papert proved mathematically that the perceptron could not be used for a class of problems associated to nonlinearly separable data distributions.

Very few investigators conducted research on ANNs during the 1970s. Albus developed his adaptive CMAC that is a distributed table-lookup system based on a view of models of human memory. In 1974, Paul Werbos originally developed the backpropagation algorithm. Its first practical application was to estimate a dynamic model, to predict nationalism and social communications. However, his work remained almost unknown in the scientific community for more than 10 years.

In the early 1980s, Hopfield introduced a recurrent-type ANN topology that was based on the Hebbian learning law. The model consisted of a set of first-order (nonlinear) differentiable equations that minimize a given energy function. In the mid-1980s, backpropagation was rediscovered by two independent groups led by Parker and Rumelhart *et al.*, as the learning algorithm of feedforward ANNs. Grossberg and Carpenter made significant contributions with the ART in the mid-1980s, based on the idea that the brain spontaneously organizes itself into recognition codes, and neurons organize themselves to tune various and specific patterns defined as SOMs. The dynamics of the network were modeled by first-order differentiable equations based on implementations of pattern clustering algorithms. Bart Kosko extended some of the ideas of Grossberg and Hopfield to develop his adaptive bidirectional associative memory. Hinton, Sejnowski, and Ackley developed the Boltzmann machine that is a modified Hopfield network that settles into solutions by a simulated annealing process as a stochastic technique. Broomhead and Lowe first introduced "RBF networks" in 1988; although the basic idea of RBF was developed before, under the name method of potential function, their work opened another frontier in NNs. Chen proposed functional-link networks (FLNs), where a nonlinear functional transform of the network inputs aimed lower computational efforts and fast convergence.

In 1988, the Defense Advanced Research Projects Agency (DARPA) listed various ANN applications, supporting the importance of such technology for commercial and industrial use. This fact triggered a lot of interest in the scientific community, which eventually led to new applications in industrial problems. Since then, the use of ANNs in sophisticated systems has skyrocketed. ANNs found widespread relevance for several different fields. Our literature review showed that practical industrial applications were reported in peer-reviewed engineering journals from as early as 1988. Extensive use has been reported in pattern recognition and classification for image and speech recognition, optimization in planning of actions, motions, and tasks and modeling, identification, and control. Figure 7.7 lists some industrial applications of ANNs, showing the most used ANN topologies and training algorithms, relating them to common fields in the industrial area. The diagram depicts a good picture of what has actually migrated from academic research to practical industrial fields.

From 2007 to 2013, there was a rapid growth of other research areas in machine learning, allowing statistical and graph approaches to be applied in recognizing fault patterns, typically consisting of data processing (feature extraction) and fault recognition, so that low-dimensional feature vectors would be mapping the information obtained in the system feature space to the fault space. Numerous AI tools or techniques have been used, including convex optimization, mathematical optimization, and classification-, statistical learning, and probability-based methods, specially, including k-nearest neighbor algorithms, Bayesian classifiers, support vector machines (SVMs), and ANNs. There has been an emergence during these years of multiagent systems (MAS) for distributed control, development, and commercialization of powerful computers, with RISC-based GPU hardware for faster computation. Add to this scenario the emergence of huge datasets provided by Internet-based company platforms, which made clear that NNs would have to be larger in size, with functional layers for feature extraction, more hidden layers for learning complex nonlinear multidimensional spaces, with adaptive clustering output layers for classification. Also, ANNs had to evolve to be capable of recursively assessing long temporal data sequences to capture dynamic behavior, as in speech recognition applications, text analysis, and weather forecasting. For the particular enhancement of smart grids and power systems, there is an important need for load/energy forecasting analysis of load, time, weather, seasons, customer behaviors, appliances, use of electric plug-in vehicles, and local energy production.

Then deep NNs evolved from the CNN, to the recurrent NN, including long-short-term-memory and gated recurrent units, the autoencoder, the deep belief network, the generative adversarial network, and deep reinforcement learning. It is established that the year 2012 marks the birth of the deep learning. Furthermore, deep learning approaches have been explored and evaluated in different application domains. Individual advanced techniques for training large-scale deep learning models to the recently developed method of deep learning models (Schmidhuber, 2015). Artificial intelligence techniques have been used successfully since 1988 for processes automation and intelligent decisions. Table 7.2 and Figure 7.8 show many supervised and unsupervised applications, and deep learning has been consistently used for different areas such as computer vision, neuroscience, biomedical engineering, and power systems, initially in smart-grid load forecasting, and more recently in deep penetration of renewable energy sources.

Energy conversion systems have two possible classes that help one to define the requirement of advanced control systems: (i) unconstrained energy systems and (ii) constrained energy systems. In reality, any energy source is constrained, because there are only finite energy resources in our nature. However, several constrained systems are modeled as unconstrained systems, in order to prevent excessive model and decision-making complexity. For example, a large power system will have several large power plants, supplying electrical power to a distribution system. The distribution company will sell that power and will care about their reliability and quality, and the users will just pay their fees and tariffs, believing that such electrical power is always available, and the electrical energy supply is not bounded. That is a simplification, but it works well in the old paradigm of centralized power plants. The constrained energy systems have a finite energy and most often a finite maximum power (which means finite maximum derivative of energy).

There are two types of constrained systems, the ones based on fossil fuel (gas, coal, oil, and hydrogen), where a certain amount of the input fuel will convert energy using a thermodynamic cycle (usually Rankine or Brayton, or a fuel cell), with inherent losses and maximum conversion efficiency, and the systems based on

Table 7.2 Activity: prediction, classification, data association and conceptualization, data filtering with neural network applications

Activity	Network topology	Network application
Prediction	• Backpropagation • Delta, bar delta • Extended delta, bar delta • Directed random search • Higher order neural networks • Self-organizing map	Use input values to predict some output, choose the best stocks in the market, predict weather identify people with health risks
Classification	• Learning vector quantization • Counterpropagation • Probabilistic neural networks	Use input values to determine the classification (e.g., is the input the letter A, is the blob of video data a plane and what kind of plane is it)
Data association	• Hopfield • Boltzmann machine • Hamming network • Bidirectional associative memory • Spatiotemporal pattern recognition	Like classification, but it also recognizes data that contain errors (e.g., not only identify the characters that were scanned but also identify when the scanner is not working properly)
Data conceptualization	Adaptive resonance network Self-organizing map	Analyze the inputs so that grouping relationships can be inferred (e.g., extract from a database the names of those most likely to buy a particular product)
Data filtering	• Recirculation	Smooth an input signal (e.g., take the noise out of a communication signal)

renewable energy (wind, solar, tidal, and geothermal). Those renewable energy systems can be sustainable, as long as the amount of energy conversion is less than the recovery of that energy by the environment. However, they are constrained because their derivative of energy should be optimized, which means there is a convex function that will define an amount of power conversion, dependent on the usage. For example, a wind turbine will have a peak power that depends on the tip-speed ratio and the output load, or a photovoltaic system will have a peak power that depends on the solar irradiation, temperature, and the equivalent impedance across its terminals.

The optimal system performance depends on coherent operation of components. For example, an engineer will understand that a compressor with heat exchangers and a throttle will make up a heat pump. But the operation of a thermodynamic system such as a heat pump requires information, measurement, and

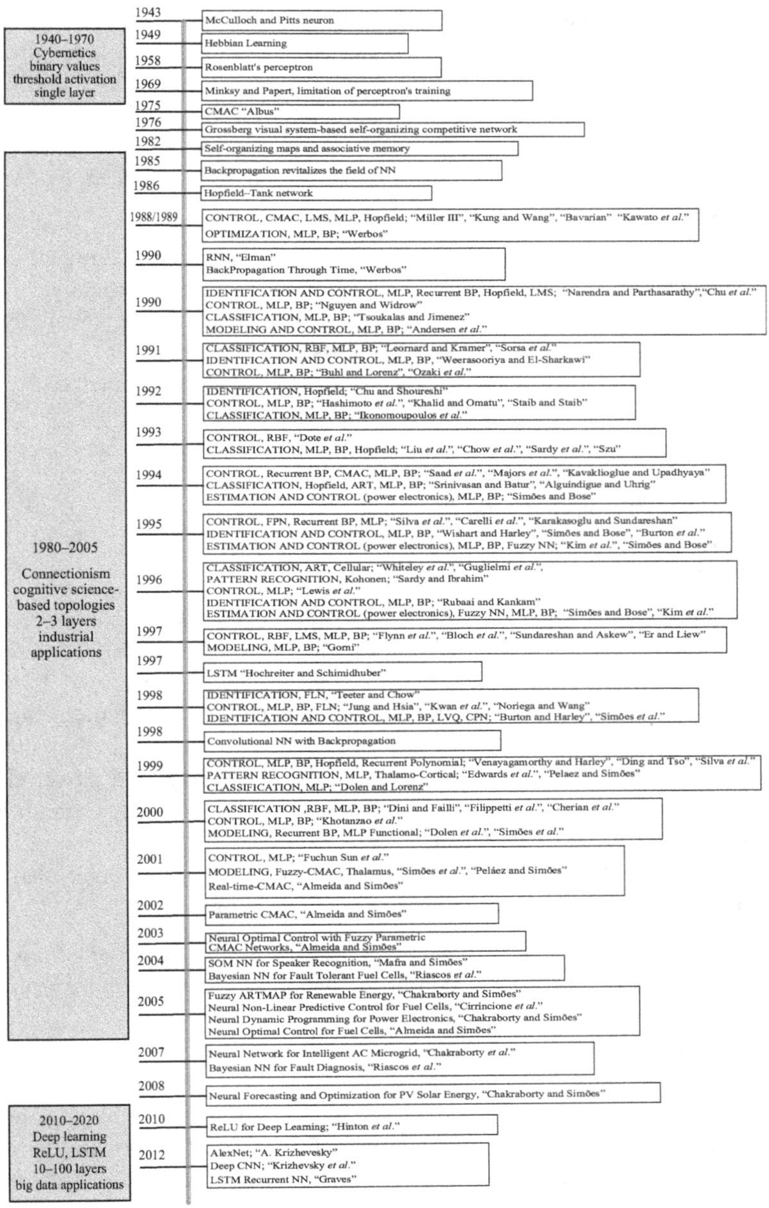

Figure 7.8 Neural network research three phases: cybernetics, connectionism, and deep learning, main advances and selected history of industrial applications

control of the compressor, which then depends on refrigerant pressure, a temperature measurement taken by a controller to evaluate how much heat is required, and a very intricate understanding of physics, chemistry, electrical, and mechanical engineering to make the heat pump to operate on its maximum efficiency. Therefore, several issues will have to be taken in consideration, and efficient energy conversion for electrical power systems will be advanced by artificial intelligence on these premises:

- parameter variation that can be compensated with designer judgment;
- processes that can be modeled linguistically but not mathematically;
- settings with the aim to improve efficiency as a matter of operator judgment;
- when the system depends on operator skills and attention;
- whenever one process parameter affects another process parameter;
- effects that cannot be attained by separate PID control;
- whenever a fuzzy controller can be used as an advisor to the human operator;
- data intensive modeling (use of parametric rules);
- parameter variation: temperature, density, impedance;
- nonlinearities, dead-band, time delay; and
- cross-dependence of input and output variables.

Chapters 6 and 7 in this book briefly introduced ANNs, approaching on backpropagation techniques and further learning topologies based on feedback, competitive, and associative paradigms. The field of NNs is very vast. Therefore, the reader is encouraged to further study other learning and training methodologies, other topologies, and learn hybridization of machine learning, computational intelligence with NNs, particularly addressing power systems and power electronics. Chapter 8 will discuss general applications of NNs and fuzzy logic in power electronics, power systems, and smart-grid technology.

In order to summarize what has been covered so far in this book, the reader should be acquainted with the following foundation ideas:

- An NN can be a mapping, or a mathematical function, capable of taking inputs and producing outputs, like a nonlinear function approximator for very complex systems.
- An NN can be represented as a computational graph, with multidimensional arrays flow, possible to implement as matrices and vectors multiplications, calculation of Euclidean distance errors, and implemented in typical computer language on any computer, or maybe in an especial FPGA, or RISC hardware device.
- An NN has layers, each of which has neurons; the most basic neuron has a weighted sum of incoming signals, going through an activation function.
- Those weights can be adapted using several types of training and convergence algorithms.
- An NN can be a static function, serving as a universal function approximator; or it can have memory of past inputs in order to implement dynamical systems; it can generate numerical outputs, classes, or clusters; or it can be used in complex engineering models for optimization and closed-loop control.

Chapter 8
Applications of fuzzy logic and neural networks in power electronics and power systems

Energy conversion systems have two possible requirements for advanced control systems: (i) unconstrained energy systems, and (ii) constrained energy systems. In reality, any energy source is constrained because there are only finite energy resources in our nature. However, several constrained systems are simplified to be unconstrained in order to prevent a very complex modeling and decision-making. For example, a large power system will have several large power plants, supplying electrical power to a distribution system. The distribution company will sell that power and will care for their reliability and quality, and the users will just pay their fees and tariffs, believing that such electrical power is always available, and the electrical energy supply is not bounded. That is a simplification, but it works well in the old paradigm of centralized power plants. The constrained energy systems have finite energy and most often finite maximum power (which means finite maximum derivative of energy). There are two types of constrained systems: the ones based on fossil fuel (gas, coal, oil, and hydrogen), in which a certain amount of the input fuel will convert energy using a thermodynamic cycle (usually Rankine or Brayton or a fuel cell), with inherent losses and maximum conversion efficiency, and the systems based on renewable energy (wind, solar, tidal, and geothermal). Those renewable energy systems can be sustainable as long as the amount of energy conversion is less than the recovery of that energy by the environment. They are constrained because their derivative of energy should be optimized, which means that there is a convex function that will define an amount of power conversion, dependent on the usage. For example, a wind turbine will have a peak power that depends on the tip-speed ratio and the output load, or a photovoltaic system will have a peak power that depends on the solar irradiation, temperature, and the equivalent impedance across its terminals.

The optimal system performance depends on the coherent operation of components; for example, an engineer will understand that a compressor with heat exchangers and a throttle will make up a heat pump. But the operation of a thermodynamic system, such as a heat pump requires information, measurement, and control of the compressor that depends on refrigerant pressure, temperature, and measurement taken by a controller to evaluate how much heat is required, and a very intricate understanding of physics, chemistry, and electrical and mechanical engineering to make such a heat pump operate on its maximum efficiency. Therefore, several issues will have to be taken into consideration, and efficient energy conversion for electrical power systems will be advanced by artificial intelligence on these premises:

- parameter variation that can be compensated with designer judgment;
- processes that can be modeled linguistically but not mathematically;
- setting with the aim to improve efficiency as a matter of operator judgment;
- when the system depends on operator skills and attention;
- whenever one process parameter affects another;
- effects that cannot be attained by separate proportional–integral–derivative (PID) control;
- whenever a fuzzy controller can be used as an advisor to the human operator;
- data-intensive modeling (use of parametric rules);
- parameter variation: temperature, density, and impedance;
- nonlinearities, dead band, and time delay; and
- cross-dependence of input and output variables.

There are typically three frameworks with some generalization of functionalities, i.e., three paradigms, that can be used for energy conversion systems, with artificial-intelligence-based computation: (i) a function approximation or input/output mapping, (ii) a negative feedback control, and (iii) a system optimization. The first one is the construction of a model, using either heuristic or numerical data, the second one is the comparison of a set point with an output that can be either measured or estimated with a function that minimizes the error of the set point with the output, and the third one is a search for parameters and system conditions that will maximize or minimize a given function. Fuzzy logic and neural network techniques make the implementation of such three paradigms possible, robust, and reliable in practical cases. The integration of modern power electronics, power systems, communications, information, and cyber technologies with a high penetration of renewable energy resources has been at the edge and at the frontier for the design and implementation of smart-grid technology. The emergence of AI techniques in past industrial applications is allowing smart-grid technology to be an interdisciplinary field with multiple dimensions of complexity. This chapter will present some background and established applications of AI in power electronics, power systems, and renewable energy systems.

8.1 Fuzzy logic and neural-network-based controller design

Conventional control has provided several methods for designing controllers for dynamic systems. All of them require a mathematical formulation for the system to be controlled, and a certain approach that will be used in order to design a closed-loop control. Some of those methods are the following:

- PID control: More than 90% of the controllers in operation today are PID controllers (or at least some form of PID controllers like a P or a PI, or an I+P controller). This approach is often viewed as simple, reliable, and easy to understand. Sometimes fuzzy controllers are used to replace PID, but it is not yet clear if there are real advantages.

- Classical control: Lead-lag compensation, Bode and Nyquist method, and root-locus design.
- State-space methods: State feedback and observers.
- Optimal control: Linear quadratic regulator, use of Pontryagin's minimum principle, or dynamic programming.
- Robust control: H2 or H∞ methods, quantitative feedback theory, and loop shaping.
- Nonlinear methods: Feedback linearization, Lyapunov redesign, sliding mode control, and backstepping.
- Adaptive control: Model reference adaptive control, self-tuning regulators, and nonlinear adaptive control.
- Stochastic control: Minimum variance control, linear-quadratic-Gaussian control, and stochastic adaptive control.
- Discrete event systems: Petri nets, supervisory control, and infinitesimal perturbation analysis.

These control approaches will have a variety of ways to utilize information from mathematical models. Most often they do not take into account certain heuristic information early in the design process but use heuristics when the controller is implemented to tune it (tuning is invariably needed, since the model used for the controller development is not perfectly accurate). Unfortunately, while using some approaches in conventional control, engineers become somewhat isolated from the control problem itself and become more involved in the mathematics; this usually allows the development of unrealistic control laws. Sometimes in conventional control, useful heuristics are ignored because they do not fit into the proper mathematical framework, and this can cause problems. Fuzzy logic and neural networks approach exactly real-life understanding instead of heavily math-oriented control by allowing heuristics and learning from past case studies or numerical data, and by retrofitting usually an excellent performance controller that most of the time excels when compared to heavily mathematical control design approaches.

An example of a control system that can be heavily mathematical oriented is the induction motor with a very complicated instantaneous model based on decoupled *d–q* equations, trigonometrical Park and Clarke transformations, in which an inverse model is resolved mathematically in order to control torque and flux with virtual *d–q* currents; then such a controller response is reverse calculated in real time in order to generate the pulse-width modulation of transistors in a three-phase inverter that commands the induction machine. It seems that fuzzy logic and neural networks are natural solutions to induction motor speed control, optimization of flux, and signal processing of nonlinear functions, i.e., the three areas described earlier. Induction generators will be further advanced in their performance with similar artificial-intelligence-enhanced modeling and control. A fuzzy logic speed control can be designed for an induction motor or DC motor drive (speed control), as depicted in Figure 8.1, in which the input signals for the fuzzy

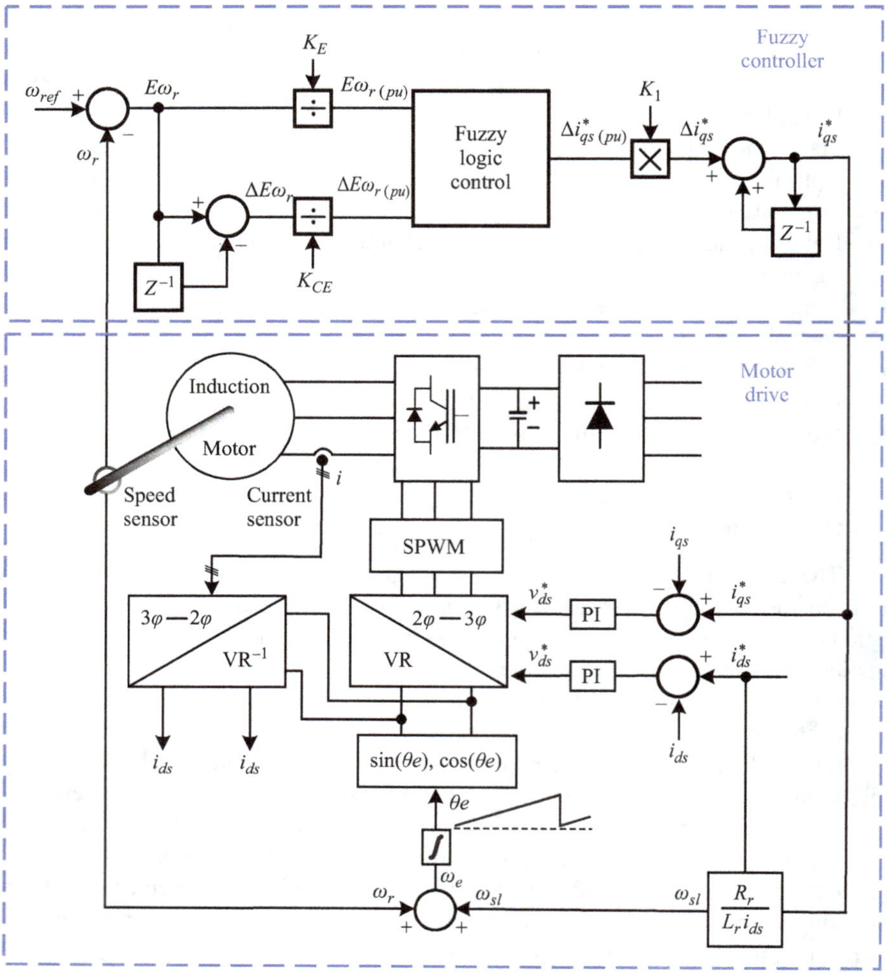

Figure 8.1 Fuzzy logic speed control system showing the input of error and change-in-error with output of the controller through accumulative summation in order to feedback the command for the electric motor. Such a fuzzy controller can also be used in other PI-like control loops

logic control are E (error) and CE (change-in-error) and the output is DU (derivative of output control). Figure 8.2 shows an indirect-vector-control-based induction generator for fuzzy logic control. Fuzzy logic control will have corresponding membership functions, where fuzzy sets are linguistically defined in Table 8.1, and Table 8.2 shows the fuzzy control rules. The universe of discourse is expressed in per-unit; such normalization allows the controller to be fine-tuned with scaling gains for E, CE, and DU. Assuming seven membership functions for each E_{pu} and CE_{pu}, there are a total of 49 possible rules. The output DU_{pu} is considered to have

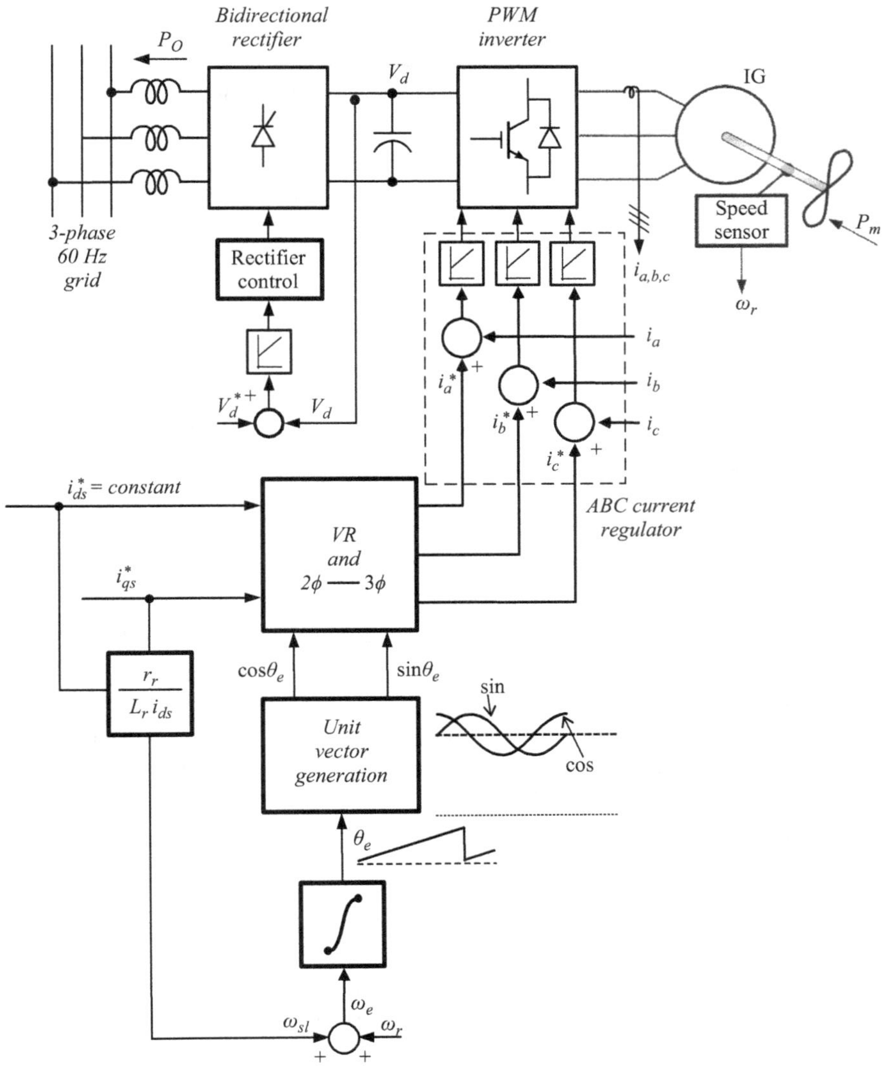

Figure 8.2 Indirect-vector-control-based induction generator for fuzzy logic control

nine membership functions. Table 8.2 shows the rule table for this fuzzy speed controller, in which the top row and the left column indicate the fuzzy sets for the variables E_{pu} and CE_{pu}, respectively, and each cell in that table gives the output variable DU_{pu} for an AND operation of those two inputs. For example, a typical rule in the matrix is like Rule 30 that reads as indicated by the following equation:

$$\textbf{\textit{IF}}\, E_{pu} = PS \textbf{ \textit{AND} } CE_{pu} = PM \textbf{ \textit{THEN} } DU_{pu} = PB \tag{8.1}$$

Table 8.1 Fuzzy linguistic variables for fuzzy control

NB	Negative big
NM	Negative medium
NS	Negative small
NVS	Negative very small
ZE	Zero
PVS	Positive very small
PS	Positive small
PM	Positive medium
PB	Positive big

Table 8.2 Fuzzy controller for a motor drive speed control loop

		Error E_{pu}						
		NB	**NM**	**NS**	**ZE**	**PS**	**PM**	**PB**
Change-in-error CE_{pu}	NB	NVB	NVB	NVB	NB	NM	NS	ZE
	NM	NVB	NVB	NB	NM	NS	ZE	PS
	NS	NVB	NB	NM	NVS	ZE	PS	PM
	ZE	NB	NM	NVS	ZE	PVS	PM	PB
	PS	NM	NS	ZE	PVS	PM	PB	PVB
	PM	NS	ZE	PS	PM	PB	PVB	PVB
	PB	ZE	PS	PM	PB	PVB	PVB	PVB

The general considerations in designing such a fuzzy controller are expressed by the following meta-rules:

1. If both E_{pu} and CE_{pu} are zero, then maintain the present control setting, i.e., $DU_{pu} = 0$.
2. If it is E_{pu} and not zero but it is approaching at a good rate, i.e., CE_{pu} has a good polarity and value, then maintain the present control setting, i.e., $DU_{pu} = 0$.
3. If E_{pu} increases, then change the control signal DU_{pu} to bring the plant output back to the conditions in which error should be zero, i.e., reverse the trend and make signal DU_{pu} dependent on the magnitude and sign of E_{pu} and CE_{pu} in order to force E_{pu} toward zero.

In several other fuzzy logic control systems, some rules are not described, either because it was found that such situations do not exist, or maybe because the output should not change. Table 8.2 depicts a fuzzy logic controller with 49 possible combinations, which can be used for a speed control of a motor drive, in which the derivative of the output must be accumulated (or integrated) in order to build up the torque control command. There are some other ways to implement a fuzzy PI, or a fuzzy PD, but this rule table mostly works well, as long as good scaling factors are retrofit for the particular application; so there are some previous simulation studies and some trial-and-error tweaking on the controller.

As discussed in Chapter 5, the rule matrix and membership functions of the variables are associated with the heuristics of general control rule operation, i.e., the meta-rules; such heuristics would be the same way by which an expert would try to control the system if he or she was in the feedback control loop himself or herself. The rules are all valid in a normalized universe of discourse, i.e., the variables are in per-unit. For a simulation-based system design, the controller tuning can be done with the MATLAB® Fuzzy Logic Toolbox, or maybe LabView is another nice environment for such design. It is also possible to develop the whole structure of the controller using C language compiled code. For advanced design, it is possible to use neural network or genetic algorithm techniques for fine-tuning the membership functions, implementing an adaptive neuro-fuzzy inference system (ANFIS). Such details are outside the scope of this chapter. This fuzzy speed control algorithm can be numerically implemented with a computer language that allows compilation into a microcontroller or a DSP.

A neural network can be used in the control of a nonlinear system as indicated in Figure 8.3. This topology is based on a model reference adaptive controller (MRAC), in which a reference model is assumed for the nonlinear plant, which can be a linearized model around an operating point of the set point. Two neural networks are used in this scheme; one of them is online learning of the inverse model of the output/input function F^{-1}. Every few cycles, after the training converges, the neural network weights of this inverse model are transferred to the neural network

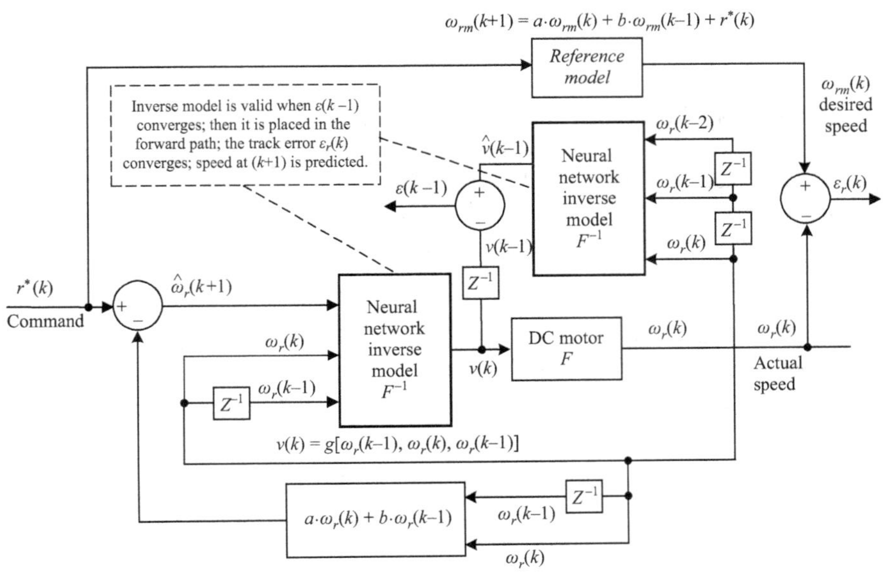

Figure 8.3 Neural-network-based inverse dynamics adaptive speed control

that controls the system. In this neural-network-based inverse dynamics adaptive control approach, the applied voltage can be considered a function of one step ahead angular speed reference $\omega_r(k+1)$, with current angular speed $\omega_r(k)$ in addition to the past one $\omega_r(k-1)$. Therefore, the following equations are applied to support such a neural network inverse model:

$$v(k) = g[\omega_r(k+1), \omega_r(k), \omega_r(k-1)] \tag{8.2}$$

$$v(k-1) = g[\omega_r(k), \omega_r(k-1), \omega_r(k-2)] \tag{8.3}$$

Figure 8.3 shows in the dashed box the neural-network-based inverse model (F^{-1}) of the machine, which is first trained off-line by the input/output data generated by the machine. The network has no dynamics, i.e., it is not of recurrent type; it can be implemented with a three-layer feedforward network with five or maybe more hidden layer neurons in which the time-delayed speed signals are generated from the actual machine speed by a delay line, or just a shift command and allocated memory. The network output signal $\widehat{v}(k-1)$ is compared to $v(k-1)$, which is generated with one-step time delay from the machine input. The training is conducted until the error signal $\varepsilon(k-1)$ becomes acceptably small. Therefore, backpropagation can be used for training, but it must be coded properly in order to use $\varepsilon(k-1)$ for convergence. Once the inverse model (F^{-1}) is trained successfully, it is placed in the forward path, by downloading the trained weights of the dashed box network into the solid line box network, which will become the inverse model control of the plant, capable of cancelling the dynamics of (F). This kind of inverse-modeling control brings an acceptable stable closed-loop control, but it might be sluggish in response. However, very complex plants could be controlled with such an approach, and the example with a DC motor was applied in 1991 as described in the study of Weerasooriya and El-Sharkawi (1991) and became an important contribution to such a control method. The system as shown in Figure 8.3 has a comparison of the actual output response with a reference model. The reference model should be a linearized version of the plant, in order to allow a tracking signal $\varepsilon_r(k)$, which should be theoretically zero, or at least bounded. If the tracking signal indicates a steady amplitude increase, it means that the plant may have parameter variation (such as thermal or saturation effects on a machine), making the need of updating the inverse model with the online training, plus an addition of a feedback loop for stability, which is a weighted summer that corrects the set point. In extreme cases, the reference model can be substituted by another average model valid in another operating point. Therefore, this MRAC neural-network-based system should have a supervisory control to make sure that the system always functions properly.

8.2 Fuzzy-logic-based function optimization

Optimization is always a concern for real-world applications, particularly when someone wants to design a system that attains the best characteristic or responds to what is an “optimal response.” Mathematically, optimization is defined as the search for a combination of parameters commonly referred to as decision variables that minimize or maximize a certain quantity, typically a scalar that is called a score or a cost, assigned by an objective function or cost function. In addition, there is a set of constraints that provide boundaries to decision variables or may define regions of infeasibility in the decision variable space. An optimization can be done off-line, such as studying all the possible configurations of a renewable energy with several sizes of wind turbines, PV arrays, batteries, hydrogen, and diesel, in order to establish the best configuration of a predesigned system that will be implemented. Otherwise, it may be a study of the best configuration for a rocket to be lifted up and reach a certain goal in the outer space. The optimization can also be required online, and typically mathematical programming has many implementation issues with online optimization. However, as told by Don Knuth, a great computer scientist,

> "... We should forget about small efficiencies, say about 97% of the time: premature optimization is the root of all evil."

Therefore, using heuristics, fuzzy logic and neural networks, an approximate optimal point can be established, and as long as this optimal condition is online and close to the real optimal solution, which is the ideal practical implementation. For example, pumping fluid will have a flow that increases but the pressure decreases, and the power for pumping initially goes up, reaching a maximum and then decreasing as a throttle continues opening. There is an optimum pump speed for the corresponding maximum value of power flow. Although a thermodynamic analysis, mass flow rate, and energetic modeling will be very complex, the idea of hill climbing can be used, i.e., a heuristic way of searching the maximum value could be based on the following meta-rule:

> —**"If the last change in the input variable (x) has caused the output variable (y) to increase, keep moving the input variable in the same direction; if it has caused the output variable to drop, move it in the opposite direction."**—such a rule can be easily implemented in any online optimization, or it can be further developed in a rule table of fuzzy statements. It can also be used to program a neural network that will learn the parameters that minimize a certain cost function using training and adaptation algorithms.

One important application of online optimization is for flux programming of induction-generator-based energy systems (Simões, 1995). Figure 8.4 depicts a fuzzy logic optimization control system in which a search of the best flux operating point and induction generator will be made on the basis of the fuzzy inference of the DC-link power generation (with inverter) and the last command of the flux quadrature current. Usually, machines operate with rated flux in order to have maximum developed torque per ampere and optimum transient response. However, renewable energy applications, such as wind or hydropower, when the mechanical shaft torque is light, and excessive flux will increase the quadrature-current component of the machine stator current, increasing the copper loss. By programming the quadrature current for a lower value, the total stator current will go down, decreasing the copper loss, and the flux will also decrease, decreasing core loss.

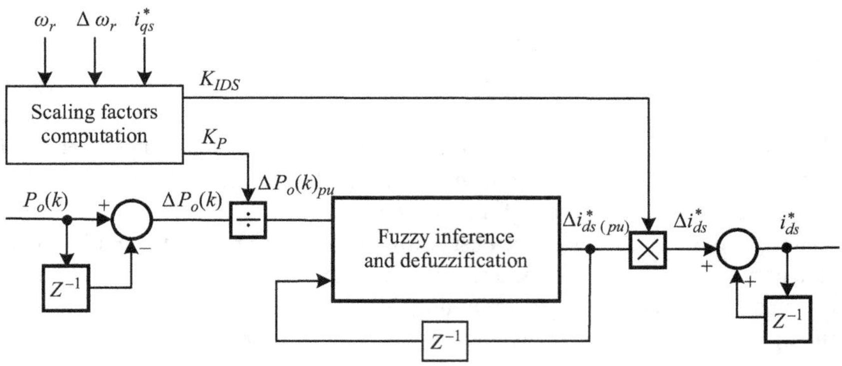

Figure 8.4 Fuzzy logic optimization control system in which a search for the best flux operating point for induction generator will be made based on the fuzzy inference of the DC-link power generation (with inverter) and the last command of the flux quadrature current

However, there is a minimum value for the flux that will maintain the system stable, so a search can be made based on heuristics: measuring the generated power, for example, at the DC link, the quadrature current is decreased as long as the generated power increases, but when the generated power starts to decrease, the flux search is reversed. Of course, a certain oscillation around the optimal point is expected, but a fuzzy logic control can be made to have adaptive large steps for the beginning of the search and progressive small steps as the best operating point is reached. The two variables should be the inputs of a fuzzy controller for flux optimization; the change (variation) of power at the DC-link $\Delta P_{d\,(pu)}(k)$ and output should be of the controller of the variation of quadrature flux current at the stator, i.e., $\Delta i^*_{ds\,(pu)}(k)$. Figure 8.5 shows the seven asymmetric triangular membership functions, comparing the variation of power $\Delta P_{d\,(pu)}(k)$ with the last variation of quadrature current, i.e., the previous one, $\Delta i^*_{ds\,(pu)}(k-1)$. Table 8.3 shows the corresponding rule table for this fuzzy controller, a typical rule reads as

$$\textbf{\textit{IF}}\Delta P_{d(pu)}(k) = \textit{Positive Small}\ (NS)\ \textbf{\textit{AND}}\ \Delta i^*_{ds(pu)}(k-1) = \textit{Negative}\ (N)$$

$$\textbf{\textit{THEN}}\ \Delta i^*_{ds\,(pu)}(k) = \textit{Negative Small}\ (NS)$$

The basic idea is that if the last control action indicates an increase of DC-link power then proceeds searching in the same direction, and the control magnitude should be somewhat proportional to the measured DC-link power change. When the control action results in a decrease of P_d , i.e., $\Delta P_d < 0$, the search direction must be reversed. At a steady state, the operation oscillates around point A, with a very small step size. The use of artificial intelligence for function optimization has been successfully used for wind and solar applications. The principles of peak power tracking control for wind energy will be discussed here, but similar principles can also be applied for photovoltaic arrays. The large energy capture of

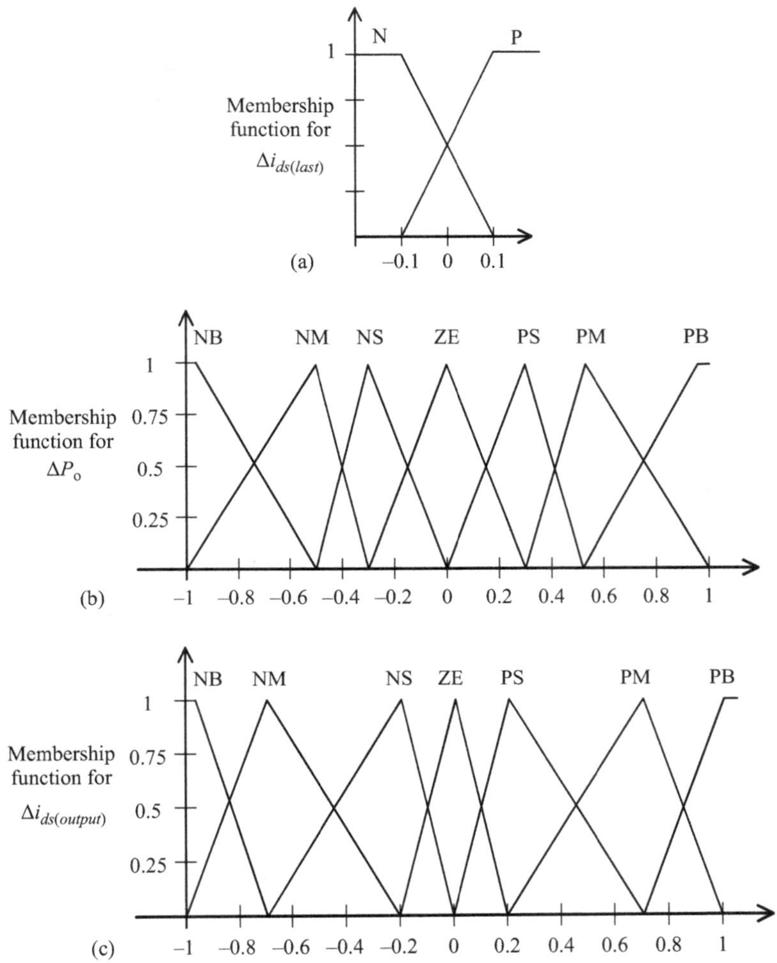

Figure 8.5 Fuzzy logic membership functions with their associate linguistic variables for change in power and change in flux quadrature current, in which the asymmetrical functions help the convergence of the searching and online optimization; (a) last variation of variation in magnetizing current, (b) last variation of converted power, (c) next setup reference for variation in magnetizing current

variable-speed wind turbines makes the life cycle cost lower, but it is required that a control system programs the wind turbine to operate at their maximum power energy conversion operating conditions. The use of artificial intelligence for function optimization has been successfully used for wind and solar applications. The principles of peak power tracking control for wind energy will be discussed here, but similar principles can also be applied for photovoltaic arrays.

The large energy capture of variable-speed wind turbines makes the life cycle cost lower, but it is required that a control system commands the wind turbine

Table 8.3 Fuzzy optimization search of best induction generator rotor flux

		Last flux change $\Delta i^{*}_{ds\,(pu)}(k-1)$	
		N	**P**
Power change $\Delta P_{d(pu)}(k)$	PB	NM	PM
	PM	NS	PS
	PS	NS	PS
	NS	PS	NS
	NM	PM	NM
	NB	PB	NB

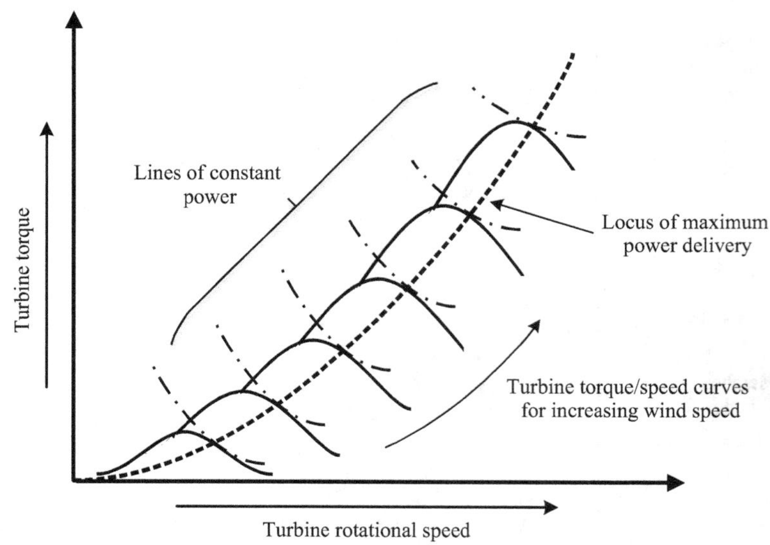

Figure 8.6 Torque-speed curves of fixed-pitch wind turbine for different wind velocities, in which the maximum power locus delivery intercepts the curves at the peak power point tracking set point

to operate at their maximum power energy conversion operating conditions. Figure 8.6 shows the torque-speed curves of a wind turbine at different wind velocities. Figure 8.9 has more details on such optimization where at a particular wind velocity (V_{W1}), if the turbine speed ω_r decreases from ω_{r1}, the developed torque increases, reaches the maximum torque at point B, and then decreases at lower ω_r. Superimposed on the family of curves, there are curves of constant power lines, indicating the points of maximum power output for each wind velocity. Therefore, as the wind velocity varies, the turbine speed should change to get the maximum power delivery, optimizing the aerodynamic efficiency; for example, in the figure for point A, or for point B, an intelligent-based controller should seek point C, which is the one that gives the maximum power conversion. Of course, a family of power curves could be plotted against the turbine rotational speed, and for that particular set of curves, the algorithm would search the apex of the curve. This figure illustrates that the peak torque for a particular wind turbine will not be necessarily the one that maximizes the power conversion.

The fuzzy logic control for optimizing the wind energy system will have an implementation block diagram depicted by Figure 8.7 with fuzzy membership functions given by Figure 8.8 and fuzzy inference rule table as Table 8.4. It is an extension of the method employed for searching the flux, with the difference that power will be maximized, instead of minimizing copper and core losses. A certain oscillation around the optimal point is expected, but a fuzzy logic control can be made to have adaptive large steps for the beginning of the search and progressive small steps, as the best operating point is reached. Two variables should be the inputs of such a fuzzy controller: the change, the output power at the grid (including the whole inverter system), P_0, i.e., for ΔP_0 positive, with the last $\Delta\omega_r^*$, we can define this variable as $L\Delta\omega_r^*$; the search is continued in the same direction. If on the other hand $+\Delta\omega_r^*$ causes $-\Delta P_0$, the direction of search must be reversed. The speed oscillates by a small increment when it reaches the optimum condition. The normalized variables $\Delta P_{0(pu)}(k)$, $\Delta\omega^*_{r(pu)}$, and $L\Delta\omega^*_{r(pu)}$ are described by membership functions as in Figure 8.8. In a search for peak power of wind turbine, there is possibly some wind vortex and torque ripple that may trap the search in a nonlocal minimum, so some amount of $L\Delta\omega^*_{r(pu)}$ is added to the current set point, similarly to a momentum factor used in a neural network. The scale factors KPO and KWR are a function of the generator speed and the turbine, and some fine-tuning might be necessary with the scaling, in order to make the system sensitive to the power variation with the turbine angular speed variation.

Figure 8.9 shows how a power search will operate; suppose the wind velocity is at V_{W4}, the power output will be at A if the generator speed is ω_{r1}; the fuzzy logic control will alter the speed in steps on the basis of an online search until reaching speed ω_{r2} in which the output power is maximum at B. If this system freezes the

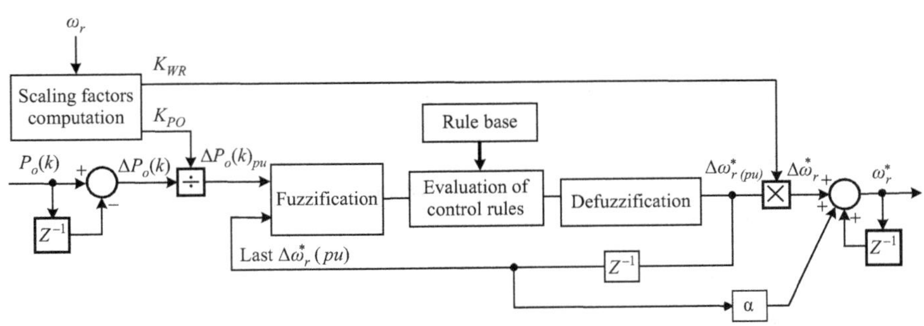

Figure 8.7 In fuzzy logic optimization control system, a search of the best rotational speed for a wind turbine is made with the fuzzy inference of the grid converter power compared to the last power output commanded by the previous rotational speed

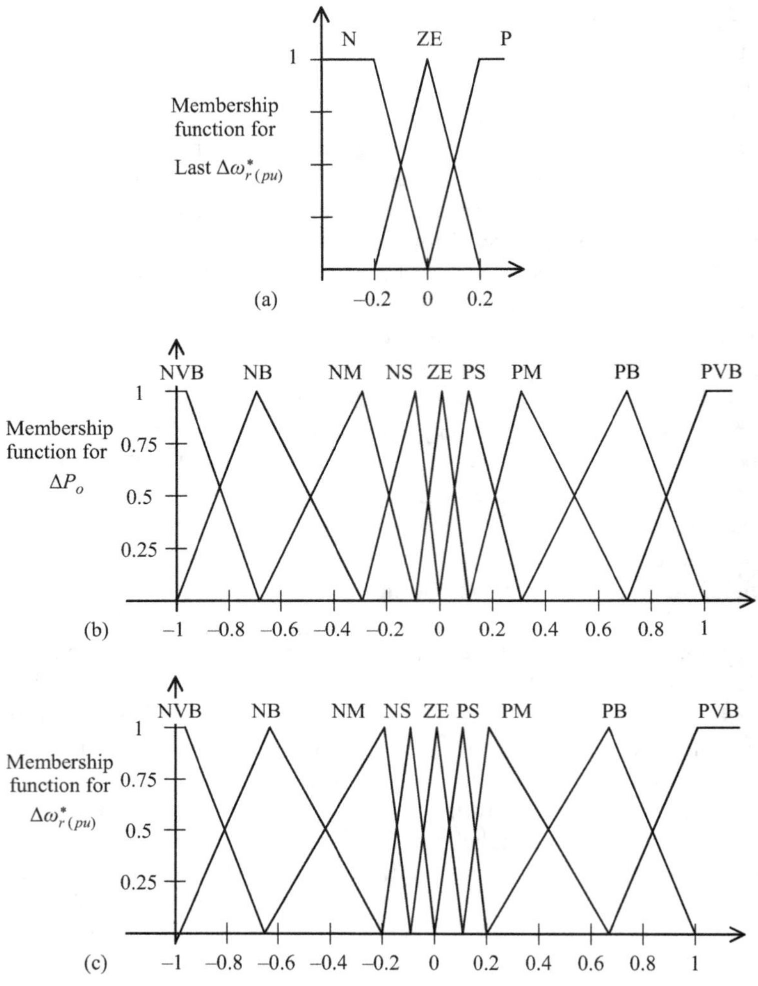

Figure 8.8 Fuzzy logic membership functions with their associated linguistic variables for change in power and turbine angular speed for searching the best peak power point of the wind turbine with online optimization. (a) last variation of variation in shaft angular speed, (b) last variation of converted power, (c) next setup reference for variation in the turbine angular speed

operating point at ω_{r2} for steady-state conditions, a next search for the best induction generator flux takes over, and the system is brought to the operating point at C. Now, if the wind velocity changes to V_{W2}, the output power will jump to point D; one fuzzy controller will bring the operating point to E by searching the speed until arriving to ω_{r4}; the system is locked on this angular speed conditions for steady state, and another fuzzy controller will search the best flux of the induction generator, bringing the operating point to F. A similar discussion can be made for a

Table 8.4 Fuzzy rules for optimization of wind turbine power for variable wind velocity

		Last speed change $\Delta\omega^{*}_{r\,(pu)}(k-1)$		
		P	**ZE**	**N**
Power change $\Delta P_{o(pu)}(k)$	PVB	PVB	PVB	NVB
	PBIG	PBIG	PVB	NBIG
	PMED	PMED	PBIG	NMED
	PSM	PSM	PMED	NSM
	ZE	ZE	ZE	ZE
	NSM	NSM	NMED	PSM
	NMED	NMED	NBIG	PMED
	NBIG	NBIG	NVB	PBIG
	NVB	NVB	NVB	PVB

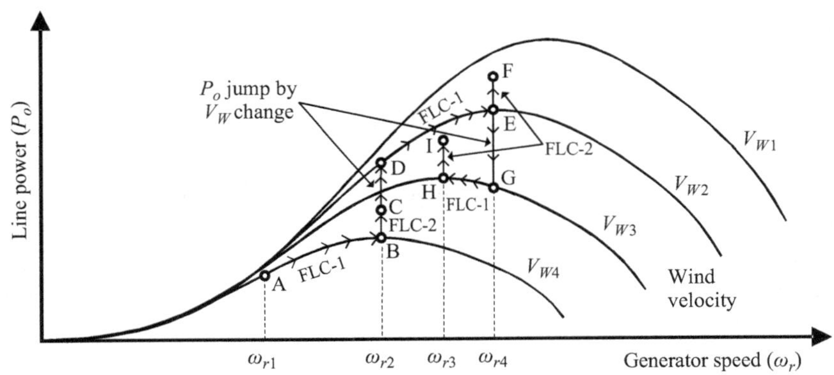

Figure 8.9 Sequential search from initial point A, FLC-1 bring to point B, with flux optimization to reach on point C; as the wind jumps the power to point D, there is another sequential optimization of FLC-1 and FLC-2 to arrive to point F, and when the wind steps down to point G, the sequential control will take to point I

decrease of wind velocity to V_{W3}. Figure 8.10 shows a complete fuzzy-logic-based control and optimization of an induction-generator-based wind turbine grid-connected energy system (Simões *et al.*, 1997b, 1998). A double PWM back-to-back voltage source converter connects the induction generator to the utility grid. The machine side converter is controlled with field oriented, or *d–q*, vector control. The torque command is impressed with a fuzzy logic controller, defined as FLC-3, which runs the same as discussed previously.

There are two online fuzzy optimization controllers: FLC-1 searches the generator speed to maximize the output power, whereas FLC-2 searches the machine excitation current to optimize generator efficiency at light power generation; those online function optimization fuzzy controllers are the same as described previously. The advantages of such fuzzy controllers are obvious: they will accept noisy and

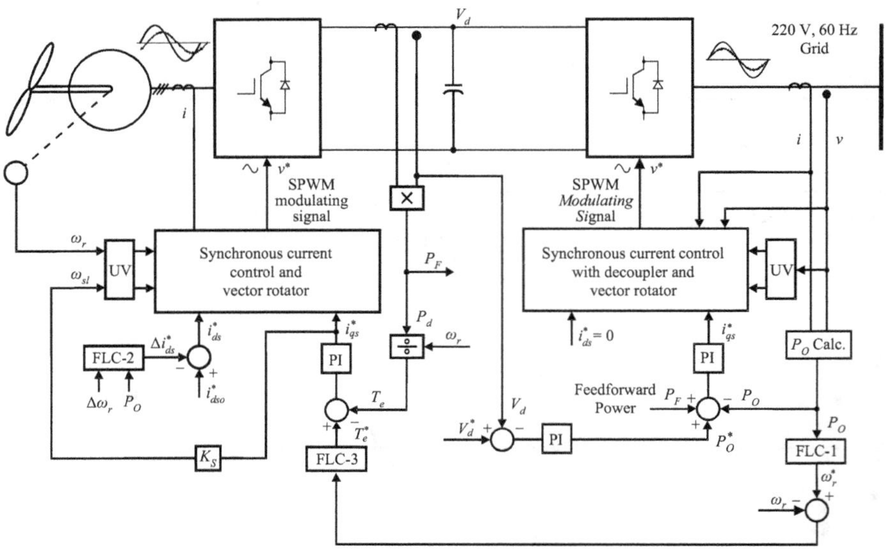

Figure 8.10 Fuzzy-logic-based control and optimization of an induction generator wind turbine grid-connected energy system

inaccurate signals; they will provide adaptively decreasing step-size in the search that leads to fast convergence; they will provide robust control on the machine shaft against wind turbine vibration and mechanical resonance; in addition, wind velocity information is not needed, and the system is insensitive to parameter variation. The principles of FLC-1, FLC-2, and FLC-3 have been tested with analysis, simulation, and implementation as given by Simões *et al.* (1997a,b), Sousa and Bose (1994), Sousa *et al.* (1995), and they can be easily translated into other kinds of applications, particularly enhancing renewable-energy-based power systems.

8.3 Fuzzy-logic-and-neural-network-based function approximation

Real-world function approximation problems are system modeling solutions, basically algebraic solutions, i.e., a mapping of input to output, or state-space solutions, i.e., memory-based equations, in which the output will depend on the internal states plus past inputs. Mathematical discussions are relevant when function approximation involves incomplete information, high dimensionality, nonlinearities, and noise. Function approximation is the problem of finding a transformation that approximates a target of another function, based on a sample of observations taken from such an unknown target. In machine learning, the function approximation formalism is used to describe general problem types in pattern recognition, classification, clustering, and curve fitting. A statistical approach will use a probability density function supporting statistical machine learning and/or density estimation, or even Bayesian algorithms. A neural network is a very simple way for easily learning a function that relates a huge dataset of input variables versus output variables. Neural networks can be used to support energy forecasting, load-flow modeling of large power systems, learning of nonlinear functions in power electronics and power systems, estimation of ill-modeled systems, for example: temperature variation effect of induction motor rotor resistance, nonlinear response of capacitors, loss modeling of transformer core, lifetime expectation of protection circuits, and so many other applications that are usually very difficult to find a function approximation using pure mathematical theory. Function approximation can be useful in several problems related to signal processing in power electronics, power systems, and power quality. One example is the estimation of distorted wave. Power converters are characterized for generating complex voltage and current waves, and it is often necessary to determine their parameters such as total rms current, fundament rms, active power, reactive power, distortion factor, and displacement factor. These parameters can be measured by electronic instrumentation (hardware and software) or estimated by mathematical model, Fast Fourier Transform (FFT) analysis, and so on. Fuzzy logic principles can be applied to fast and reasonable accurate estimation of those parameters, due to their enhanced nonlinear mapping (or pattern recognition) property. In Simões and Bose (1993), a fuzzy-logic-based pattern recognition was applied for the first time in power electronics, in which the estimation of a diode rectifier line current wave was discussed and studied with simulation analysis of two methodologies, that are, the Mamdani method and the Takagi–Sugeno approach. Comprehensive details can be read in the study of Simões (1995), but the main idea is to observe the pulsed nonlinear current waveforms and to use the width (*W*) and height (*H*) for each pulse. For a single-phase rectifier, there is one pulse per semi-cycle of voltage line, while for the three-phase rectifier, for each semi-cycle of phase voltage, the line rectifier current has two pulses. While using the Mamdani method, or Type 1, several rules can be designed as

$$\textbf{\textit{IF}}\ H = PMS\ \textbf{\textit{AND}}\ W = PSB\ \textbf{\textit{THEN}}\ I_s = PMM,\ I_f = PSB\ and\ DPF = PMS$$

where the power factor will be numerically calculated as $PF = DPF(I_f/I_s)$, i.e., each rule gives multiple outputs. In Simões and Bose (1993), the development and the accuracy of a fuzzy TS, or Type 2, estimation is also compared, in which a rule would read as

$$\textbf{\textit{IF}}\ H = PMS\ \textbf{\textit{AND}}\ W = PSB$$

$$\textbf{\textit{THEN}}\ I_s = a_0 + a_1 H + a_2 W,\ I_f = b_0 + b_1 H + b_2 W\ and\ DPF$$

$$= c_0 + c_1 H + c_2 W$$

where the linear coefficients can be found with numerical examples based on experimental data, fitted with the least-square method. It is obvious that Type 2 approach is more precise and has a more compact rule table than Type 1. Detailed information and discussion is available in the studies of Simões (1995) and Simões and Bose (1993).

Neural networks are very powerful algorithms for real-world approximation problems, which can be algebraic solutions or state-space ones. Power electronics, power systems, and power quality have several signal processing phenomena involving incomplete information, high dimensionality, nonlinearities, and noise. Neural networks have been applied to various control, identification, and estimation in power electronics and drives; some of these applications are as follows:

- single- or multidimensional lookup table functions,
- converter PWM,
- neural adaptive PI drive control,
- delay-less filtering,
- vector rotation and inverse rotation in vector control,
- drive MRAC,
- drive feedback signal estimation,
- online diagnostics,
- estimation of distorted waves, and
- FFT signature analysis of waves.

Figure 8.11 shows the block diagram of a direct vector-controlled induction motor drive, with a feedforward neural-network-based estimator, which was originally published as the first application of neural network in power electronics as given in the studies of Simões (1995) and Simões and Bose (1995).

A backpropagation neural network has been trained to estimate rotor flux (ψ_r), unit vector $(\cos\theta_e, \sin\theta_e)$, and torque (T_e) by solving (8.4)–(8.11). A DSP-based estimator was used in Simões and Bose (1995) for comparison with the neural network. Such a network is feedforward trained with backpropagation, with instantaneous mapping; so the machine terminal voltages were initially integrated by a hardware low-pass filter in order to generate the stator flux signals. These variable

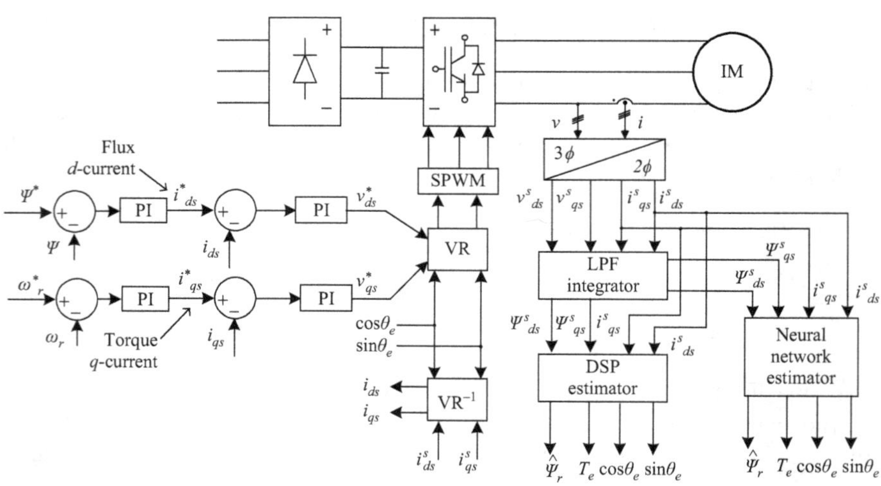

Figure 8.11 Modern adjustable speed vector control with neural network estimation

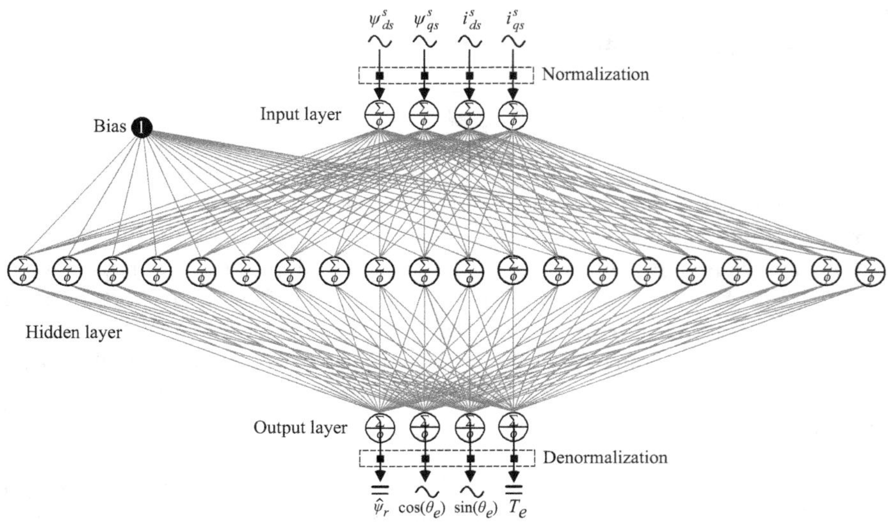

Figure 8.12 Neural network for estimation of vector control motor drive signals

frequencies and variable magnitude sinusoidal signals have been used to calculate the output parameters, with a topology with three layers, in which the hidden layer has 20 neurons, as indicated in Figure 8.12.

The input layer neurons have linear activation features, but the hidden and output layers have a hyperbolic tangent-type activation function, in order to allow bipolar outputs. This network is capable of correctly and accurately tracking torque, flux, and unit vector signals that were tested with high- and low-inverter frequencies, working satisfactorily for closing the loop of an adjustable speed modern induction motor drive.

$$\psi^s_{dm} = \psi^s_{ds} - i^s_{ds} L_{ls} \tag{8.4}$$

$$\psi^s_{qm} = \psi^s_{qs} - i^s_{qs} L_{ls} \tag{8.5}$$

$$\psi^s_{dr} = \frac{L_r}{L_m}\psi^s_{dm} - L_{lr} i^s_{ds} \tag{8.6}$$

$$\psi^s_{qr} = \frac{L_r}{L_m}\psi^s_{qm} - L_{lr} i^s_{qs} \tag{8.7}$$

$$\widehat{\psi}_r = \sqrt{\left(\psi^s_{dr}\right)^2 + \left(\psi^s_{qr}\right)^2} \tag{8.8}$$

$$\cos\theta_e = \frac{\psi^s_{dr}}{\widehat{\psi}_r} \tag{8.9}$$

$$\sin\theta_e = \frac{\psi^s_{qr}}{\widehat{\psi}_r} \tag{8.10}$$

$$T_e = \frac{3}{2}\left(\frac{P}{2}\right)\left(\psi^s_{dr} i^s_{qs} - \psi^s_{qr} i^s_{ds}\right) \tag{8.11}$$

8.4 Neuro-fuzzy ANFIS—adaptive neural fuzzy inference system

The term, neuro-fuzzy systems or neuro-fuzzy methods or models, refers to a combination of techniques from neural networks and fuzzy systems, in which a fuzzy system is typically created from data by some kind of heuristic learning procedure used in neural networks. Those systems are usually used for function approximation problems. The manual design of a fuzzy system requires a structure (set of rules) that is usually built on trial and error and expertise from operational knowledge of the system. One of the first combinations of neural network learning method with concepts of fuzzy systems was published in Keller and Hunt (1985), in which a perceptron learning algorithm for classification was enhanced with a fuzzy membership of data items in the searched classes in order to improve the convergence of the learning algorithm.

ANFISs are a class of networks that are functionally equivalent to fuzzy inference systems, which build a powerful intelligent system with improved design and performance features. As discussed earlier, the membership functions and rule table of a fuzzy inference system would be based on the experience of the operator, or designer, or expert with the heuristics of that particular system. It was also discussed that Type 2, or Takagi–Sugeno–Kang inference, would have consequents with linear functions that could have their linear coefficients trained by least squares (off-line) or recursive least squares (online). In 1990, Horikawa proposed, for the first time (Horikawa *et al.*, 1990), a hybrid technique that would allow a neural network structure to resemble a Type 2 fuzzy inference, and backpropagation could be used to fine-tune the membership functions and even find relevant rules of inference, based on sufficient input/output data for a process. The literature discusses both zero- and first-order Sugeno methods. As an example, a zero-order ANFIS can assume X and Y as input variables, where F is the defuzzified output signal, with A_1, A_2, B_1, and B_2 supposed to be triangular membership functions for the antecedent part of the rules, and f_1, and f_2 are assumed to be output singleton MF, as indicated in Figure 8.13. The two rules to be considered in this example are

Rule 1 ***IF*** X *is* A_1 ***AND*** Y *is* B_1 ***THEN*** $Z_1 = f_1$

Rule 2 ***IF*** X *is* A_2 ***AND*** Y *is* B_2 ***THEN*** $Z_2 = f_2$

The output Z can be constructed as (8.12) in which W_1 and W_2 are the degrees of truth for Rule 1 and Rule 2. The ANFIS architecture as depicted in Figure 8.13 can be implemented using any compiled or script language, or in MATLAB, or using the fuzzy logic toolbox of MATLAB.

$$Z = \frac{W_1 f_1 + W_2 f_2}{W_1 + W_2} = \frac{W_1}{W_1 + W_2} f_1 + \frac{W_2}{W_1 + W_2} f_2 \tag{8.12}$$

The feedforward neural network will have five layers, with specific connections so as to resemble the calculations inside the fuzzy inference Type 2, so the degree of

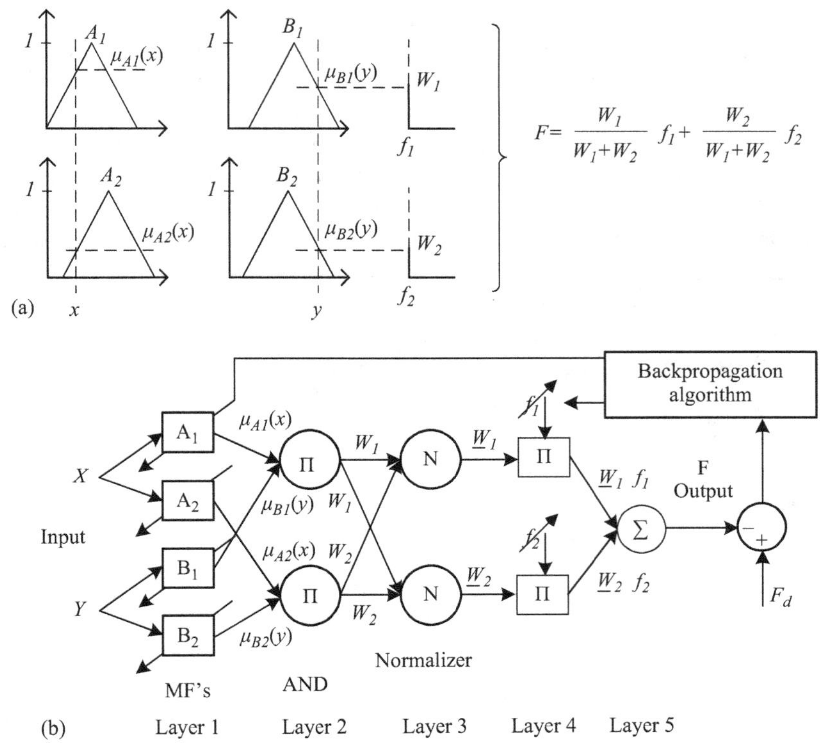

Figure 8.13 ANFIS simple system: (a) zero-order Sugeno fuzzy inference system, (b) ANFIS structure with backpropagation

truth for each rule can be calculated by multiplication of the evaluation of membership functions, and the firing strengths can be normalized. The outputs can be calculated by multiplying with the consequent of the parameters, and then each contribution of each rule is summed (since they are normalized). The calculated output F of the network is compared with the desired value F_d and the error signal is used to train the network parameters by backpropagation algorithm. The MATLAB fuzzy logic toolbox can be used to design such an ANFIS, with several types of membership functions such as Gaussian, sigmoidal, and so on. Further reading on this topic is available on the studies of Kim *et al.* (1996) and SOBRAEP (n.d.).

For the interested readers, there is a suggestion for a practical ANFIS exercise: let us assume a system with one input $u(k)$ defined with five membership functions, in which the output relations are given by linear equations defined by five rules:

Rule 1 **IF** $u(k)$ is *VERY SMALL* **THEN** $y_1(k) = a_1u(k) + b_1u(k-1)$
Rule 2 **IF** $u(k)$ is *SMALL* **THEN** $y_2(k) = a_2u(k) + b_2u(k-1)$
Rule 3 **IF** $u(k)$ is *MEDIUM* **THEN** $y_3(k) = a_3u(k) + b_3u(k-1)$
Rule 4 **IF** $u(k)$ is *LARGE* **THEN** $y_4(k) = a_4u(k) + b_4u(k-1)$
Rule 5 **IF** $u(k)$ is *VERY LARGE* **THEN** $y_5(k) = a_5u(k) + b_5u(k-1)$

where the input is $u(k)$ and the output is defined by segmented outputs, $y_1(k), y_2(k), y_3(k), y_4(k)$, and $y_5(k)$ depending on a fuzzy reasoning for input $u(k)$ to be very small, small, medium, large, or very large. Each rule will have an evaluation for the output that depends on current input and past inputs with linear coefficients. Assume that the membership function has smooth shapes, given by the exponentials, as given by (8.13)–(8.17) where M_{vs}, M_s, M_m, M_l, and M_{vl} are parameters to define the proper membership functions. Explain the role of each weight, draw an ANFIS network, and find a way to calculate $w_i (i = 1, 2, 3, 4, 5)$ for a given set of input/output patterns. Then, using ANFIS with the MATLAB fuzzy logic toolbox, implement your system in order to compare with your hand evaluation, in which the output of the system is given by (8.18). The whole system, from input to output, can be defined by equations that are differentiable, and backpropagation can be derived for training the weights, i.e., the coefficients for the five consequents (THEN) linear equations in the five rules.

$$\mu_{\text{Very Small}}(u) = \exp\left(-\frac{(u - M_{vs})^2}{0.5^2}\right) \tag{8.13}$$

$$\mu_{\text{Small}}(u) = \exp\left(-\frac{(u - M_s)^2}{0.5^2}\right) \tag{8.14}$$

$$\mu_{\text{Medium}}(u) = \exp\left(-\frac{(u - M_m)^2}{0.5^2}\right) \tag{8.15}$$

$$\mu_{\text{Large}}(u) = \exp\left(-\frac{(u - M_l)^2}{0.5^2}\right) \tag{8.16}$$

$$\mu_{\text{VeryLarge}}(u) = \exp\left(-\frac{(u - M_{vl})^2}{0.5^2}\right) \tag{8.17}$$

$$y = \sum_{i=1}^{5} w_i y_i \tag{8.18}$$

Such an example for a simple ANFIS can be integrated with a large number of inputs. The outputs for each rule could be trained to give, for example, a severity of a faulty, based on historical data for training. Then a hybrid model could be used, in which the probability of those faults would be incorporated in a Bayesian network and both outputs, from the ANFIS (giving the severity), and the Bayesian network (giving the probability), would be concatenated into a risk matrix, assuming a linear matrix model indicated by the following equation:

$$[risk]_{n\times 1} = [probability]_{n\times m} \times [severity]_{m\times 1} \tag{8.19}$$

8.5 AI-based control systems for smarter power systems

Artificial intelligence, particularly fuzzy logic and neural networks, can be applied to renewable energy systems and smart grid; the current literature (Bose, 2017a) just shows their potential in the past few years. However, they can be summarized in the following applications:

- consumer load forecasting on the grid,
- forecasting of wind and PV generation curves,
- online fault diagnostics and fault-tolerant control on power system,
- sensorless robust estimation of feedback signals,
- noise and delayless filtering of signals,
- neural network modeling of static and dynamical system elements and real-time simulation by DSPs/FPGAs/ASIC chips,
- intelligent scheduling of generation and storage,
- high-performance intelligent control of system elements, and
- real-time pricing predictions of electricity with demand-side management.

In a wind farm, the overall system could be designed for intelligent monitoring and protection. In such an application, the signals to be acquired should be as follows:

- Wind signals: velocity, wind direction, and turbulence;
- Turbine signals: blade speed, shaft speed, pitch angle, pitch angle control signal, bearing temperatures, vibration of blade, yaw angle, shaft torque, mechanical brake signal, and tip-speed-ratio;
- Gear box: oil temperature, oil viscosity, noise intensity, vibration, and nacelle temperature;
- Generator: bearing temperatures, shaft vibration, stator winding temperature distribution, rotor magnet temperatures, shaft torque, stator voltages, phase sequence, percentage of terminal voltages and currents imbalance, stator currents RMS, average, peak, stator frequency, active power, and reactive power;
- Converter: converter temperatures, cooling fluid velocity, DC-link voltage, DC-link current, DC-link power, AC line voltages, output frequency, phase unbalance of voltages, AC line currents, phase unbalance of currents, active power, reactive power, and motoring/regeneration mode; and
- Fourier and wavelet expansion of selected signals.

The signals can be monitored with the help of sensors, or adaptive sensorless estimation, to determine the general health condition of a wind farm, such as indicated in Figure 8.14. The health conditions could be "excellent" if the variation of the signals remains confined to a highly satisfactory range. If some signals go beyond this range, but are yet very safe, the system can be defined as "very good." Similarly, for other ranges, the health index can be classified as "good," "fair," "poor," "unsafe," etc. If some signals degrade, the diagnostic messages for the signals can be generated independently. If any signal goes beyond the safe range, for a faulty condition, the system can be shut down for protection. Similar health-monitoring principles can be

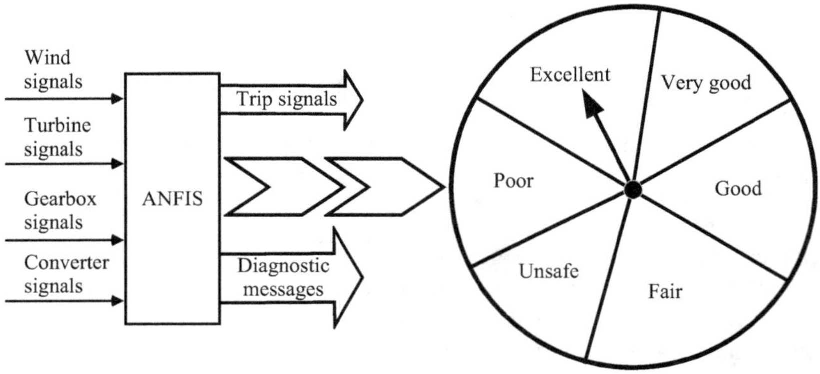

Figure 8.14 AI-based wind-farm health monitoring system

extended to PV or other systems, eventually implemented in a real-time smart-grid platform (such as Opal-RT or other possible solutions).

8.6 Artificial intelligence for control systems

The application of artificial intelligence for control systems is related to the function approximation approach. Chapter 5 discussed the fuzzy methodology, as depicted in Figure 5.2, in which the operator is the model to be identified. Therefore, an artificial neural network (ANN)-based controller will also learn the system dynamics to handle the system. While designing a suitable ANN-based model, the choice of computing implementation is very important. Either a compiled or script language can be used for implementing an NN; MATLAB, LabView, and Modelica allow powerful implementation with toolboxes already available; there are some shell-based software for neural network design, with simple drag-and-drop commands and data analysis. In addition to MATLAB, Mathematica, and Maple, there are other tools available for the development of neural networks, such as Python that is an interpreted, high-level, general-purpose programming language. Its design philosophy emphasizes code readability with its notable use of significant white space. Its language constructs and object-oriented approach aim to help programmers write clear, logical code for small and large-scale projects; also, Julia is a high-level dynamic language with a surface similarity to MATLAB. In the last few years, the resurgence of neural network research has been rebranded as deep learning (as discussed in Chapter 9 of this book), with some tools developed by Internet-company-based corporations, for example, PyTorch is an open source machine learning library based on the Torch library, used for applications such as computer vision and natural language processing, developed by Facebook's AI Research lab, and TensorFlow that is a free and open-source software library for dataflow and differentiable programming across a range of tasks. It is a symbolic math library and is also used for machine learning applications such as neural networks, developed for both research and own production at Google. However, it is not necessary to be a computer scientist or an expert in any specific high-tech language or coding, to work with most of the paradigms in neural networks, as long as the main principles and fundamentals are understood, and the reader considers a learn-by-doing path toward their final implementation.

Adaptive plant modeling is the identification of the plant in which a neural network will match the plant for a given input signal and the minimization of an error function. Another important procedure is the inverse plant identification in which the neural network would be matched the plant for a given input spectrum; if such an adaptive inverse is cascaded with the plant, such as deconvolution, the path in the closed-loop transfer function will behave as a gain. In the mathematical perspective, if the plant has poles and zeros, making an inverse would be a problem only for a non-minimum phase plant, because such an inverse would have unstable poles, but the time span of such a neural network could be made adequately long, so the mean square of the error of the optimized inverse would be made for a small fraction of the plant input power density distribution. Then the inverse is achieved with proper delayed set points for optimized transport delay and compensation of the unstable poles. Neural networks have been applied very successfully in the identification and control of dynamic systems. The universal approximation capabilities of the multilayer perceptron make it a popular choice for modeling nonlinear systems and for implementing general-purpose nonlinear controllers. Figure 8.15 shows the principles of a time-series approach, in which delayed time-window inputs are sent for a backpropagation neural network, i.e., the network learns the past time-window transfer function, saving it for the next one; Figure 8.15 shows in (a) a time window must be selected in accordance to the system dynamics, and in (b) a feedforward neural network is fed with tapped delay line, current and past (N-1) sampled input.

In the system identification stage, a neural network model of the plant to be controlled is developed. In the control design stage, such a neural network plant model will be used to train a controller; three possible architectures can be used after the system identification: (i) model predictive control, (ii) adaptive inverse model-based control, and (iii) model reference control. In the model predictive control, the plant model is used to predict the future behavior of the plant, and an optimization algorithm is used to select the control input that optimizes future performance. Figure 8.16 shows in (a) how the prediction error between the plant and neural network output can be used as training signal, and in (b) a feedforward neural network tapped delay line; such a scheme in which the prediction error between the plant output and the neural network output is used as the neural network training signal. One of the most commonly used feedback NNs is NARX. It is a recurrent network with feedback connections enclosing several layers of the network. The NARX network has many applications, which can be used for modeling nonlinear dynamic systems and can also be employed for nonlinear filtering purposes to make the target output as a noise-free version of the input signal, as indicated in Figure 8.17 where the adaptive inverse model based control happens continuously (after training, the follower network copy the weights, frozen for

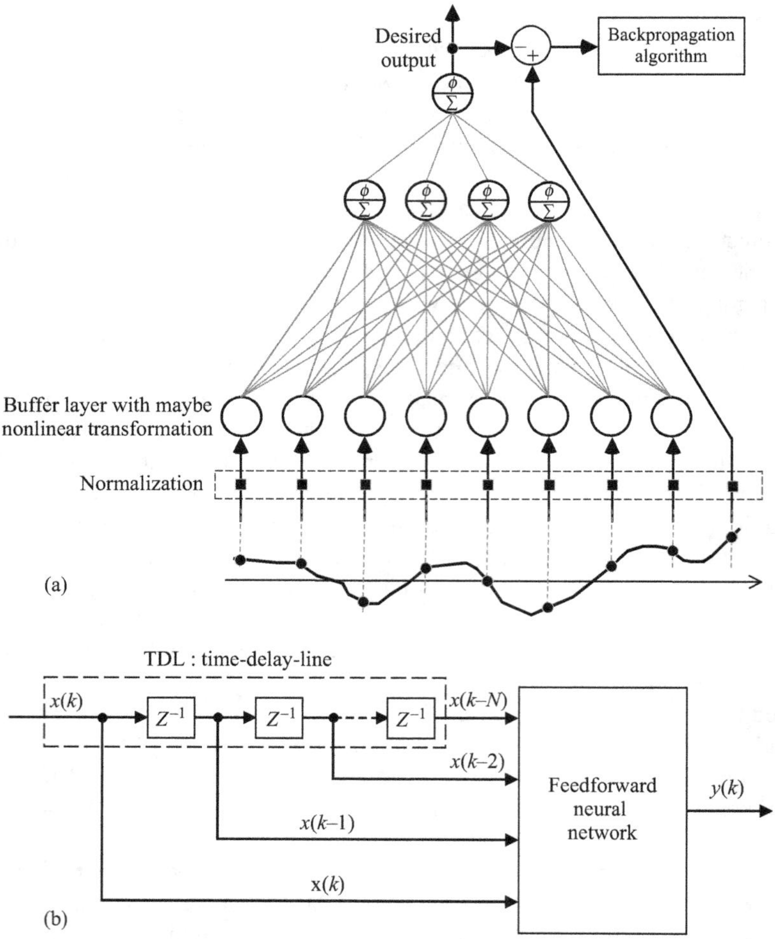

Figure 8.15 Time-delayed inputs for neural network function approximation: (a) time window must be selected in accordance with the system dynamics, (b) a feedforward neural network is fed with tapped delay line, current, and past (N−1) sampled inputs

inverse control application). After a cycle of complete training, the Slave network copies the weights, which are frozen for the inverse controller, but the inverse model training continues to ensure robustness against parameter variation. Figure 8.18 shows the model reference control; the controller is a neural network that is trained to control a plant so that it follows a reference model. The neural network plant model is used to assist in the controller training, i.e., the plant model is identified first, and then the controller is trained so that the plant output follows the reference model output. There are three sets of controller inputs: (i) delayed reference inputs, (ii) delayed controller outputs, and (iii) delayed plant outputs.

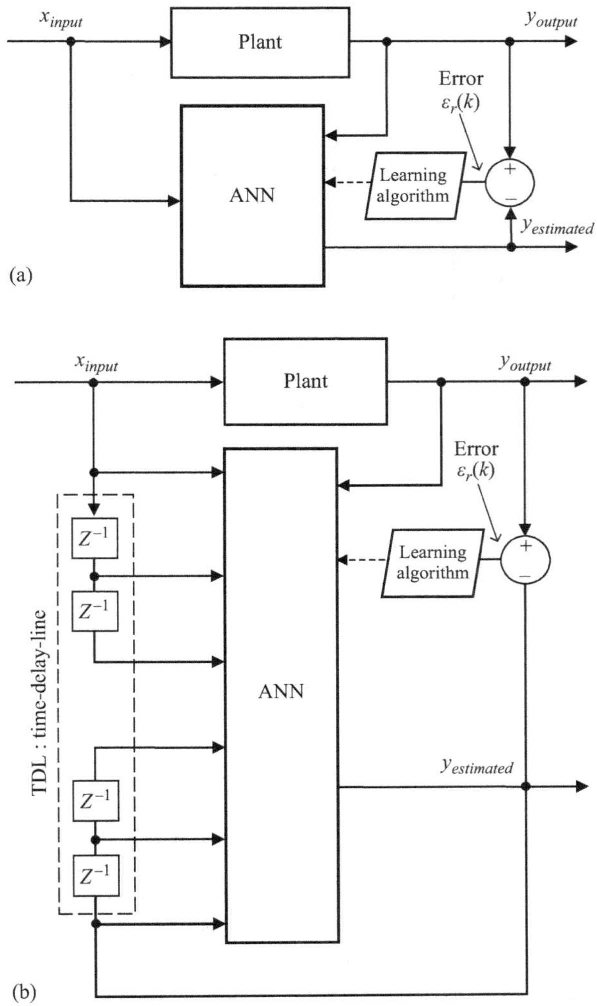

Figure 8.16 Model predictive control: (a) prediction error between the plant and the neural network output used as a training signal, (b) a feedforward neural network tapped delay line

For each of these inputs, you can select the number of delayed values to use. Typically, the number of delays increases with the order of the plant. There are two sets of inputs to the neural network plant model: (i) delayed controller outputs and (ii) delayed plant outputs.

It has been a constant challenge for researchers to find optimal AI-based solutions to design, manufacture, develop, and operate new generations of industrial systems as efficiently, reliably, and durably as possible. Getting enough

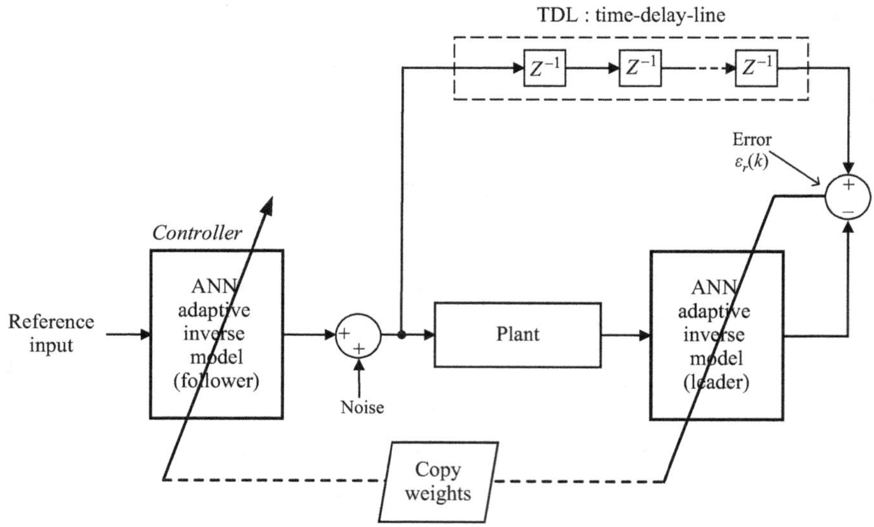

Figure 8.17 Adaptive inverse model-based control, in which an input perturbation or pseudorandom noise is introduced, so the adaptive process occurs continuously. After training, the follower network copies the weights, frozen for inverse control application

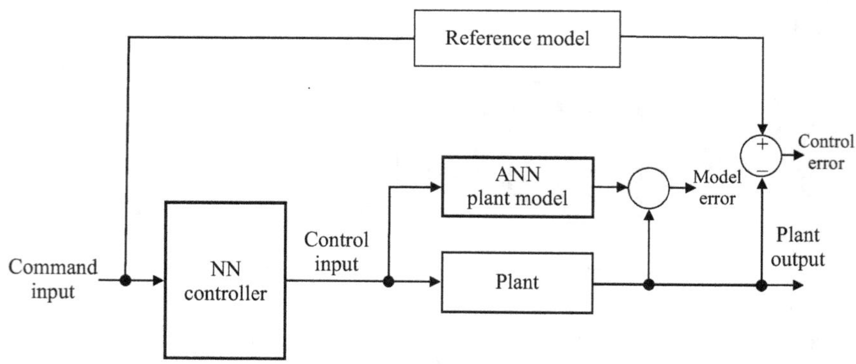

Figure 8.18 Model reference control, NN block diagram implementation

information about the system, that is, to be modeled, is the first step in system identification and modeling process. Besides, a clear statement of the modeling objectives is necessary for making an efficient model. Industrial systems may be modeled for condition monitoring, fault detection and diagnosis, sensor validation, system identification or design, and optimization of control systems. Both fuzzy logic and ANNs have the computational power to solve many complex problems; it can be used for function fitting, approximation, pattern recognition, clustering, image matching, classification, feature extraction, noise reduction, extrapolation (based on historical data), and dynamic modeling and prediction.

There are so many more potential applications for smart-grid power electronics and power systems (Bose, 2019a). Particularly for data intensive applications, ANN-based model building procedure includes system analysis, data acquisition and preparation, network architecture, as well as network training and validation. Applications and limitations of ANN approach for system identification and modeling were also discussed in this chapter. It is important to notice that approximation and error are inseparable parts of any system identification method, and ANN is not an exception. Many issues should be considered when a comparison is made between ANN and any of the conventional modeling techniques. Neural networks can be applied to control and signal processing in power electronics. Considering the input–output property of a feedforward network, it can be used in one or multidimensional function mapping. The training by backpropagation appears to be a lookup table like implementation, but the network can interpolate between the example data values. A typical application is the selected harmonic elimination method of PWM control in which the notch angles of a wave can be generated by a neural network for a given modulation index. A neural network can receive time-delayed inputs of a distorted wave and perform harmonic filtering without phase shift. Although a feedforward network cannot incorporate any dynamics within it, a nonlinear dynamical system can be simulated by time-delayed input, output, and feedback signals. The problem of estimation of parameters of distorted waveforms with a neural network has been addressed, in which the calculation of rms current, fundamental rms current, displacement power factor, and power factor for varying firing angle, load impedance, and impedance angle is performed in a topology with two hidden layers. A current control PWM scheme has been designed in which the network receives the phase current error signals through a scaling gain and generates the PWM logic signals for driving the inverter devices. The transfer function is the clamp type, and the output signals have eight possible states corresponding to the eight states of the inverter. An inverse dynamic neural network-based model controls DC drive with a nonlinear load allowing unknown dynamics of the motor and the load that are captured by a feedforward neural network. The trained network is then combined with the reference model to achieve the trajectory control of speed. A three-layer network with five hidden layer neurons was trained off-line to emulate the inverse model of the machine. Since the reference model is asymptotically stable and the tracking error tends to be zero, the network is placed in the forward path to cancel the plant dynamics. The same structure can also be extended to AC drives. A flux observer for induction motor drives has been explored, in which the neural observer receives the synchronously rotating frame stator voltages and currents, then it estimates the rotor flux magnitude and unit vectors for vector transformation. In addition, there is a rotor time constant identification unit that helps fine-tune the estimator. The observer consists of two emulator units, i.e., a neural flux emulator and a neural stator emulator. The flux emulator receives the delayed rotor flux at the input and calculates the slip frequency and the rotor flux at the output. The neural stator emulator receives the flux, frequency, and stator currents at the input and generates the stator signal at the output. After the off-line training is finished, the online training is executed with the rotor time constant identification unit. A flux estimator for field-oriented control of an induction motor has been implemented in which a neural network is trained with start-up input raw data; further details are in Bose (2006, 2017b, 2019b) and Bose (2002).

Chapter 9

Deep learning and big data applications in electrical power systems

Power systems are massive and complex electrical engineering systems. In the past few years, with the advent of the smart-grid paradigm, we have been witnessing a high penetration of wind and solar power with an active participation of customers. They may buy and also sell electrical power, and for this reason they are usually called prosumers.

Power systems used to function only under unidirectional power flow, i.e., power would flow from the generation, to the transmission, to the distribution, reaching the final users. Of course, some power flow capability at the transmission level has been implemented since the initial design of any multi-regional grid. However, at the distribution level the bidirectionality and multi-functionality of power flow is a new feature, implemented in the past few years.

The power system analysis and decision-making has been dependent only on physical modeling, numerical calculations, and some statistical inferences. Contemporary smart grids have bidirectional power flow, and uncertainty on the random nature of renewable energy availability, also a geographical dispersion of mobile loads (such as hybrid electric vehicles), with a partial observability of power-quality issues. A new generation of power-electronics-enabled power systems hardware, electrical circuits instrumentation, communications, intelligent control, and real-time performance is shaping the present and future development of smart-grids. Engineers must develop the technology for smarter power systems in order to build smart-grids, and big data applications are a requirement for such modernization.

Modern distribution installations have widespread advanced metering infrastructures (AMI), wide area monitoring systems, and other monitoring/management systems. Massive data are available that can be used for model development and training in artificial intelligence (AI) applications. Deep learning methods and architectures are powerful tools to improve solar and wind generation prediction accuracy, based on large datasets, providing effective solutions for managing flexible sources, load forecasting, scheduling, and net-metering transactions. Demand response (DR) allows customers to shift their load from peak periods to off-peak periods and to decrease their electricity usage during peak time. Smart meters provide data that reflect users' energy consumption behavior. Such data can support load decomposition and price forecasting, allowing consumers to make right decisions, where DR can be successfully implemented. The future of the electric utility industry and the growth of smart-grid systems will have power-quality applications based on data science calculations and transformations on such big datasets.

The US National Institute of Standards and Technology (NIST) supported a working group called Big Data Working Group to establish a common terminology for big data analysis (BDA), where they focus attributes of those datasets to parameters such as volume, velocity, variety, and variability.

- *Data volume* refers to the amount, or quantity, of data. Data increase exponentially for the electric power industry (Ausmus *et al.*, 2019). Modern electric grids have AMI for collecting data; there are about 130 million meters in the USA (as of this time, Spring 2021), and the increased number of clients with AMI are contributing for a large data volume. Increasing deployment of PQ meters and advanced protective relays also contribute to data records.
- *Data velocity* is the rate in which data are transmitted and received. In power systems, an example of equipment with high data velocity is a phasor measurement unit (PMU). Legacy PMUs have sampling rates up to one sample per second. However, future PMUs are expected to have multiple sample rates per second, in accordance with C37.118.1-2011—IEEE Standard for Synchrophasor Measurements for Power Systems (Zobaa *et al.*, 2018). Power-quality meters have high sampling rates to observe current and voltage waveforms, usually 128 samples per cycle. The future smart-grid distribution system with AI may require even higher sampling rates.
- *Data variety* refers to a diversity of sources used to measure data. For instance, modern electric grids monitor temperature for various devices. Power transformers, transmission and distribution line conductors, and capacitor banks can have their life cycle, replacement strategy, and asset management based on many measurements such as (i) gas pressures, (ii) temperature, (iii) humidity, and (iv) dielectric parameters. Breakers and relays may have several status operating conditions. It is possible to have several other sensors and remote monitoring. Such a universe of possibilities increases the variety of power system data.
- *Data variability* describes the changes in a dataset rather than a change in the individual measurement. Data variability is related to the need for dynamic scaling to efficiently handle the additional computational processing. As an example, dynamic scaling may have variable sampling for PMUs and PQ meters because a continuous sampling rate may have lower resolution, but when a transient occurs, a higher sampling rate is required to properly capture the dynamics of the event.
- *Data value* has been defined by the IEEE Smart Grid Big Data Group (Gadde *et al.*, 2016), which refers to gaining useful information or "value" out of a dataset using data science (Chang, 2015); in accordance with NIST, the definition of "value" is a fundamental data-science learning feature.

Conventional time-domain model simulations are time-consuming as well as very challenging when implementing large-scale nonlinear models. It is necessary to have precise knowledge of model parameters, for thousands of generators, transmission lines, and loads (Nabavi and Chakrabortty, 2013). Conventional model-based methods also require results of measurements, plus model-based reduction combining aggregation theory with system identification. BDA may help deriving reduced-order modeling in order to predict the sensitivity of the power flow oscillations inside any area with respect to faults as well as wind and solar power penetration in sections of the distribution grid. Utilities could exploit this information from simulations of the aggregated model and evaluate their dynamic coupling with neighboring companies, leading to more efficient resource planning.

The analysis of transmission and distribution has been traditionally conducted as completely decoupled infrastructures, in which the design engineer will select a section and apply the Thévenin-equivalent model for simplifying the modeling attributes. However, using the Thévenin equivalent model implicates that overall efficiencies from sources to the loads are not considered, and they are particularly for renewable energy applications, where maximum peak power tracking must be considered for overall energy conversion.

Integrated co-simulation and analysis on both transmission and distribution sides of the grid can be conducted using machine learning (ML) and other AI methods. Aggregation techniques (de Souza *et al.*, 2019) can be used with new instantaneous power theories (Harirchi *et al.*, 2019; Harirchi and Simões, 2018; Simões *et al.*, 2019) and advanced signal processing. Deep learning models can be used in updating generator and load setpoints, based on the load forecasts, as well as incorporating online estimation algorithms about weather, storage, net-metering, possibilities of natural disasters, approximate generation schedules over the next hour, and information about transmission line faults that may severely disrupt the currents flowing on the distribution side. Load control algorithms can be tuned whether the distribution grid depends on the transmission grid or on dispersed generation.

9.1 Big data analytics, data science, engineering, and power quality

Figure 9.1 illustrates several reasons to apply big data analytics in power systems, concerning the data volume for smart-grid systems. There is data accumulation on power quality in exponential proportions for monitoring higher order harmonics. Equation (9.1) shows that any instantaneous time-domain signal (current) is composed of their fundamental components plus a residual component, containing DC, integer, and non-integer harmonics.

Alternating periodic non-sinusoidal instantaneous electrical current may be defined as

$$i(t) = i_f(t) + i_{res}(t) \tag{9.1}$$

where i_f is the fundamental component and i_{res} is defined as residual component, which contains DC, integer, and non-integer multiple of the fundamental component. The IEEE Standard 1159 discusses that current harmonics in a 60-Hz AC

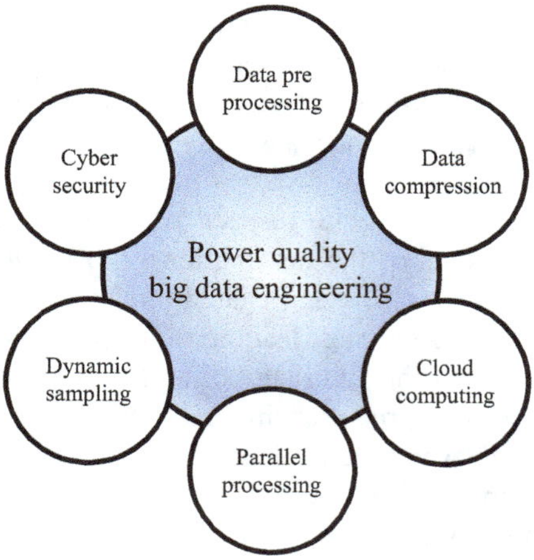

Figure 9.1 Tools for a big data analytics as a framework for power-quality applications

system should be monitored in the range from 0 to 9 kHz (150th harmonic) (Bhattarai *et al.*, 2019). So, the current can be expressed by

$$i_h(t) = \sqrt{2}\sum_{h=2}^{N} I_h \sin(h\omega t + \beta_h) \tag{9.2}$$

where the subscript h represents the order of harmonics, $i_h(t)$ is the total current with all harmonic components, β_h is a phase shift angle for h-harmonic, I_h is the rms value of the h-order harmonic. In practice, it is common to have industrial PQ monitoring devices that only monitor current harmonics up to the 50th, and in such a case the current is be given by

$$i_h(t) = \sqrt{2}\sum_{h=2}^{50} I_h \sin(h\omega t + \beta_h) \tag{9.3}$$

Nonlinear loads such as variable-frequency drives, light emitting diodes (LED), and modern electronics devices generate harmonics in the order of 2–9 kHz; electric vehicles (EV) and inverter-based distributed energy resources (DERs) inject high-order current harmonics in the same order as well. Therefore, there is a need for modern PQ monitoring devices to increase the range of current harmonic monitoring up to the 150th order. In addition to current harmonics, many modern electric loads generate harmonics in the range of 9–150 kHz, called "supra-harmonics." In that case, the current will be given by (9.4), where s is the order of supra-harmonics:

$$i_{sh}(t) = \sqrt{2}\sum_{s=151}^{2,500} I_s \sin(s\omega t + \beta_s) \tag{9.4}$$

Electrical power systems may also have inter-harmonics (non-integer multiples of the fundamental) that may be defined as

$$i_{ih}(t) = \sqrt{2}\sum_{i \neq h}^{150} I_i \sin(i\omega t + \beta_i) \tag{9.5}$$

where i is the order of inter-harmonics. Equation (9.1) can be written as the expanded format indicated next, assuming there is no DC component:

$$i(t) = i_f(t) + i_{res}(t) = i_f(t) + i_h(t) + i_{ih}(t) + i_{sh}(t)$$

$$i(t) = i_f + \sqrt{2}\sum_{h=2}^{150} I_h \sin(h\omega t + \beta_h) + \sqrt{2}\sum_{s=151}^{2,500} I_s \sin(s\omega t + \beta_s)$$

$$+ \sqrt{2}\sum_{i \neq h}^{150} I_i \sin(i\omega t + \beta_i) \tag{9.6}$$

where $i_f = \sqrt{2}I_1 \sin(\omega t + \beta_1)$. In this equation it is considered harmonics for a high penetration level of nonlinear loads, such as the ones produced by power electronics of EVs and DERs. Equation (8.6) shows that it is required a considerable higher sampling rate, leading to a significant increase in both data velocity and volume for PQ meter devices. In conclusion, in smart-grid systems BDA plays an important role in assessing power-quality monitoring and control of distributed generation, with a high penetration of renewable energy and modern loads, requiring advanced data storage and real-time data retrieval, complex mathematical signal processing, and AI for enhanced decision-making.

9.2 Big data for smart-grid control

Figure 9.2 shows the evolution of the power grid, initially from an unidirectional power flow, then a second stage (which are in most parts of the world) showing the lead-follower technology, as well as a third topology with bidirectional power, integrated communications, and advanced infrastructure. In the current state-of-the-art, i.e., the second stage in Figure 9.2, power system controllers are designed for damping transient stability and power oscillation phenomena. There is some operation in a decentralized fashion using local output feedback. Master–slave approaches with a distribution control center allow demand-side control and optimized generation for mostly static load profiles. However, the rapid modernization of the grid needs advanced distributed control and tight communication integration. System-wide-coordinated controllers become essential where signals measured at one part of the network are communicated to remote parts for feedback; such a paradigm is considered wide-area-control (WAC). Some utilities may use graph-sparse based algorithms for optimization and implementing controllers that require a lesser number of communication links still keeping good closed-loop performance. In advanced

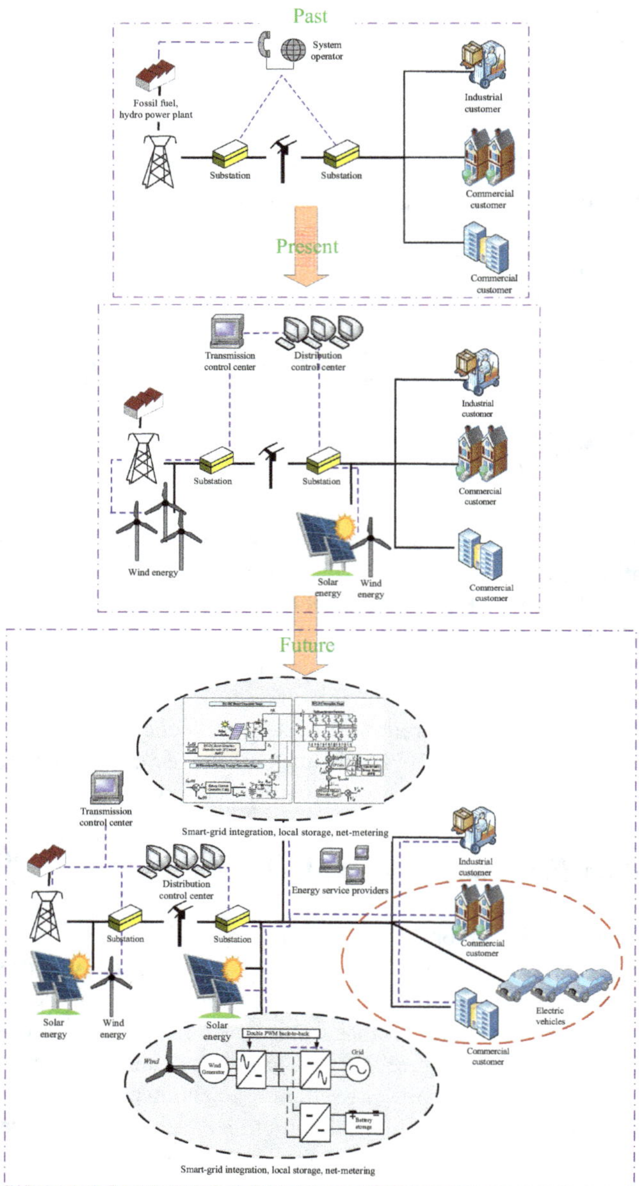

Figure 9.2 Evolution of the power-system, from unidirectional power flow, current technology, and future bidirectional power with integrated communications and advanced infrastructure

applications, it is possible to develop a fully distributed control based on multi-agent systems, even incorporating AI, making the present systems more toward the third stage shown in Figure 9.2, for such inverter-based resources. The traditional power system has a legacy on voltage/frequency control mechanisms (droop control), and new mechanisms for inverter-based DERs have been developed (Laudahn *et al.*, 2016). During the last few years, AI and advanced signal processing has been incorporated into inverter-based resources (Babakmehr *et al.*, 2020). In recent years a technique named phase-space method has been used for islanding detection and classification, such as the mathematical approach where the data of a time series are reconstructed within a higher dimensional space, as described in Eguiluz *et al.* (2000) and Ji *et al.* (2011) and later in Harirchi *et al.* (2019), Khamis *et al.* (2018), Khamis and Shareef (2013). Figure 9.1 shows tools and potential solutions in the framework for power-quality applications approaching big data engineering, i.e., handling large amounts of data (Chang, 2015):

- *Data acquisition and preprocessing:* Before data are used, some preprocessing of it is necessary, and maybe some data are missing or inaccurate; data cleaning techniques are used to mitigate such errors, interpolation, and resampling.
- *Data compression:* In order to retain the necessary information and also reduce the size of the original dataset, lossy compression or lossless compression techniques can be used. In lossless design, the compression does not lead to any loss of features in data. In lossy compression, some information may be removed. On the other hand, lossy compression may increase network performance in terms of bandwidth, throughput, and latency, and also optimizing data storage resources (Harb and Jaoude, 2018; Babakmehr *et al.*, 2016).
- *Parallel processing:* Parallel computing is splitting a complex computational problem into smaller ones, to be solved by individual processors, or by individual cores in a large computational group of processors. There are specific computer languages made for parallel processing (Tsai *et al.*, 2016).
- *Cloud computing:* A distributed computer networked system for enabling ubiquitous, on-demand network access, where a shared pool of computing resources, including data storage, remote processing, and remote applications, can be provisioned through service provider interaction (Tsai *et al.*, 2014).
- *Dynamic scaling/sampling*: Dynamic sampling is a technique for scaling down the sampling time of a monitoring device when it is at steady state, but increasing sampling rate for higher in change of states events, and transient response.
- *Cyber security:* As the electric grid and PQ evolve into a big data and multi-dimensional space, there is an increased interdependency across power and cyber technologies, with potential risks for cyber-attacks (Carvalho *et al.*, 2018). Therefore, it is very important that future power-quality applications have cybersecurity measures to identify and mitigate cyber threats.

The power grid is evolving by shifting energy supply from large central generating stations to smaller DERs. Therefore, power electronics technology is the enabler.

Most signal-processing-based feature extraction methods are not instantaneous—most often voltage and current signals are recorded for a window of time and stored—such a data is processed through a signal decomposition to calculate the corresponding Fourier or wavelet coefficients (features) for further analysis. Such a computational processing methodology imposes a time delay, maybe affecting bandwidth and real-time control parameters. On the other hand, each processing cycle could be calculated separately. For example, coupling information between the three-phase fault behavior could be missing in the case of imbalanced conditions and calculations made on per-phase model. Therefore, in the past few years instantaneous power theories allowing feature extraction and classification of power-quality events have been under development for enhancing smart-grid with real-time high performance control (Harirchi and Simões, 2018; Simões *et al.*, 2019; Simões and Bubshait, 2019; Oruganti *et al.*, 2019; Brandao *et al.*, 2019; Busarello and Pomilio, 2015; Bubshait and Simões, 2018; Busarello *et al.*, 2018; Mortezaei *et al.*, 2018; Ansari and Simões, 2017; Bubshait *et al.*, 2017; Yang *et al.*, 2016; Simões *et al.*, 2016).

9.3 Online monitoring of diverse time scale fault events for non-intentional islanding

Smart-grid systems require comprehensive awareness and status online monitoring, including instantaneous fault detection and localization. When a portion of the power system (mainly in the distribution level) is disconnected from the main grid that remains energized, and there is no cascading black-out failure, such an operation is defined as non-intentional islanding. In general, a DER based on renewable micro-generation, such as PV or wind sources, may provide flow of power to these sub-systems, with local energy backup such as a battery system. Once an outage happens in the power grid, the energy flow can be maintained in the system (supported from micro-generators). Of course, constraints must be considered for reconnection, allowing perfect synchronization of the DER with the main grid. Some typical power-quality issues are encountered on some typical faulty scenarios. There are many possible timescales, and fault events could be (i) voltage sag, (ii) harmonic, transient, and (iii) oscillations, or a combination of those events. Islanding might be triggered depending on the fault duration and depth. The following events are the most common three-phase fault scenarios in distribution systems:

1. capacitor switching (possible power-quality issues: transient oscillatory, harmonic);
2. arc furnace (possible power-quality issues: flicker);
3. induction motor start-up (possible power-quality issues: sag, flicker);
4. lightening (possible power-quality issues: impulse, swell);
5. line to ground (possible power-quality issues: voltage fluctuation, sag, harmonic, unbalance, and swell);
6. line to line (possible power-quality issues: voltage fluctuation, sag, harmonic, unbalance);

7. single-phase faults for nonlinear load (possible power-quality issues: sag, harmonic, unbalance);
8. three-phase faults for nonlinear load (possible power-quality issues: harmonic);
9. three-phase faults for power-quality issues, such as a three-phase sag;
10. transformer energizing (possible power-quality issues: harmonic, unbalance).

9.4 Smart electrical power systems and deep learning features

Typical ML models can be trained using either *supervised* or *unsupervised learning*. Supervised learning is widely adopted in (i) classification and (ii) regression tasks, whereas unsupervised learning is the most common choice for (iii) clustering tasks, as shown in Figure 9.3. Either supervised or unsupervised learning could have an artificial neural network (ANN) trained on several kinds of learning algorithms.

Reinforcement learning (RL) is yet another paradigm for training ML models, which cannot be considered supervised or unsupervised. It has four basic components: an agent, the environment the agent interacts with, the actions taken by the agent to interfere on the environment, and a reward function that models the quality of the action results. The objective of RL is for the agent to maximize the cumulative reward by taking a series of actions in response to a dynamic environment. RL can be used to train models for sequential decision tasks under uncertainty. A typical RL algorithm operates with only limited knowledge of the environment and with limited feedback on the quality of the decisions.

When neural networks (NNs) became widespread, utilized in several engineering and computer science applications by the middle of the 1980s, it was

Machine learning (ML)		
Supervised learning		Unsupervised learning
Classification	Regression	Clustering
Support vector machines (SVM)	Linear regression (GLM)	*K*-Means
Decision trees	Support vector regression (SVR)	Hierarchical
Naive Bayes	Ensemble methods	Gaussian mixture
Nearest neighbor	Decision trees	Hidden Markov model
Neural network	Neural network	Neural network

Figure 9.3 Machine learning algorithms by functionality

accepted that NNs should be shallow, *which* means just one hidden layer, or maximum two hidden layers. The main reason for such limitation is that the backpropagation training method, used in most supervised learning tasks, suffers from the problem of *vanishing gradients* (Hochreiter *et al.*, 2001). Backpropagation computes the gradient of a loss function with respect to the NN weights (parameters), based on the chain rule of calculus, which involves the cumulative multiplication of gradient terms. As the error signal from the output layer is propagated back through the hidden layers to the input, there is an exponential decrease of the resulting gradient product to well less than 1. In other words, the early layers either train very slowly or do not move away from their random starting positions; input layers are very important, because they detect features. The age of deep learning started about 2005–10, when greedy layer-wise pre-training based on auto encoder structure started to be implemented in multiple layers NN topologies, accelerating the training. In addition, new activation functions were introduced to tackle the vanishing gradient problem, such as rectified linear units (ReLUs), with very great performance achieved with convolutional and recurrent networks. Figure 9.3 shows some capabilities implemented on most of power systems with AI capabilities, based on ML functionalities:

1. *Classification*: The objective is to predict categorical labels of new input data based on past classifications from historical data. For instance, historical patterns of AMI-hacked data and healthy AMI data could be used in binary classification to predict whether a smart meter has been hacked. It is also possible that data might be classified into more than two possibilities, which is called a multiclass classification.
2. *Regression*: Usually based on statistical analysis, this method uses historical data input to develop a model to predict one or more output variables. Regression methods are used for forecasting load, weather conditions, renewable energy generation, power system optimization of generation, and load profiles, as well as electricity pricing in dynamic energy markets.
3. *Clustering*: Clustering techniques organize data into subgroups or clusters. One example of clustering used in a power system is load profiling clustering for electricity pricing. Power quality can use clustering techniques for load disaggregation based on electrical power signal signatures and pattern recognition (de Souza *et al.*, 2019).
4. *Summarization*: This method provides a compact description for a subset of data, i.e., when there are redundant variables, summarization techniques could be used to reduce the amount of data in both transmission and storage, alleviating big data issues.
5. *Association*: Describes dependencies and association relationships among different attributes. There are many variables in power systems *that* may be correlated to specific outcomes, for example, the impact of forecasted weather on the system demand for the next-day generation and load profile.
6. *Sequence analysis*: Focuses on finding sequential patterns in datasets. This could be useful for the analysis of cascade failures to identify critical assets to the electric grid.

With the increasing high penetration of solar and wind power in the electric grid, evolving bidirectional power, mobile prosumers (such as HEVs), integrated communications, and advanced infrastructure (depicted in Figure 9.2), the scheduling and operation of smarter power systems are compromised with challenges of uncertainty, random generation, and mobile flexible loads. Accurate forecasting of energy demands at different echelons in an integrated power system is very important for reliability and resilience. There are a multitude of methods for electricity load forecasting, most of them are based on datasets of specific areas, utilities, and customized studies. Near future smart-meters and cognitive-meters will provide a tremendous opportunity with pervasive and massive data that will be useful for deep learning algorithms. Therefore, it is very important to understand functional, statistical, and geometrical learning approaches for AI in smart-grid technology.

9.5 Classification, regression, and clustering with neural networks

Most applications of ML are approached as a classification activity. For example, a medical dataset might be used to classify who has diabetes (positive class) and who does not have the disease (negative class). Another example would be a financial dataset to classify if customers will default on their credit (positive class) against a group of customers who will not (negative class). Social networks may use a diversity of data that would rank a watching list of film suggestions, or music tracks to listen, or maybe classifying people on a liberal versus a conservative political profile. A database of students who took a test might be classified in pass/fail, or maybe supply several classes identified as letter grades (A, B, B+, C, C+, D), or same database could provide a continuous output for a numerical grade, e.g., from 0 to 100 points. Therefore, a continuous output could be designed with a function approximator methodology, or mapping, or sorting the output in bins, and those bins would be output classes. In electrical power systems we may have a dataset of behavior of loads or appliances in residential and commercial installations, and a smart implementation could be for nonintrusive appliance load monitoring or nonintrusive load monitoring (NILM), in order to enhance advanced metering instrumentation in smart-meters. For such AI application, a *classifier*, such as depicted in the block diagram of Figure 9.4, would need a "training dataset."

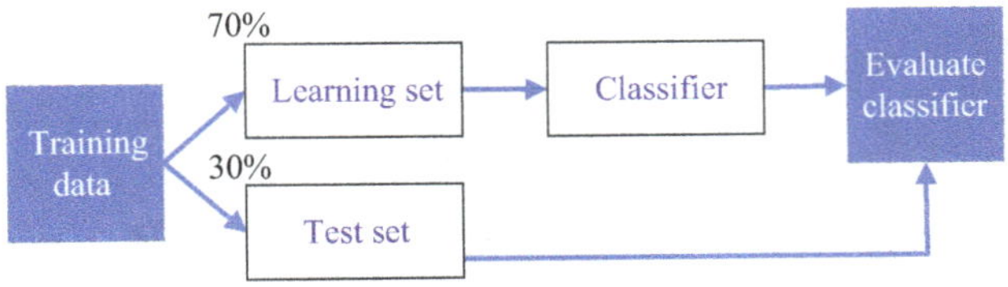

Figure 9.4 General classifier block diagram

After such a *classifier* is trained, defined by convergence parameters or by a learning index, we could assume that such a *classifier* will have learned the data distribution of the training dataset. Therefore, the system would be ready, i.e., the *classifier* could be used with new data, never presented during the training phase (test set). If the learning of data distribution has been properly achieved, the *classifier* would predict the class probabilities of the new data with a good accuracy and reliability, provided that the test set data distribution is a good approximation of the production environment. There are three typical data classifiers based on different mathematical approaches:

1. *Regression*, this is an algebraic function, with a mathematical formulation of the dataset, i.e., it maps the input values correctly to the output values, this approach is excellent when there are equations based on physics, chemistry, or biology understanding of the process.
2. *Naive Bayes classifier*, it is modeled after a statistical (Bayesian) approach, and decision-making can be implemented with a Bayesian tree, where the Bayesian (conditional) probabilities are calculated directly by creating a hash table containing the probability distributions of the features; it is generally not very highly accurate, but it can be modeled with basic statistical analysis.
3. *Geometrical approach classifiers*, they are trained as a perceptron, or a group of perceptrons, or SVM (support vector machines) methodology. This type of classifier is trained using an NN-based topology.

Figure 9.5 shows how classifiers can be implemented by (1), (2), and (3) discussed earlier. The geometrical approach, where a line, or a plane, or a hyperplane separates one or more classes, in their feature space is the one considered to support a deep learning approach, when multidimensional data, with great data volume, high data velocity, increased data variety, deep data variability, and broad data value are supposed to be used for data signal classification, using NN topologies. AI is a whole domain of computational systems that can perform tasks and activities normally considered to require human intelligence. A subgroup of AI is ML that involves the development and implementation of algorithms to enable a computer to extract, or learn, functions, mappings, and clusters from a dataset. Within ML, for the geometrical approach, there is the field of NNs, extremely vast,

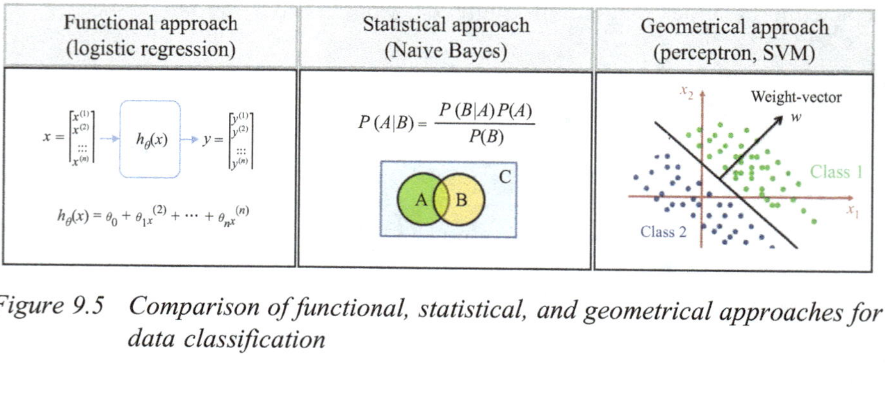

Figure 9.5 Comparison of functional, statistical, and geometrical approaches for data classification

with many topologies and training algorithms. Deep learning is just one technique that uses specific topologies of NNs, considered in the current days as one data science or data engineering field by itself.

ML algorithms can be categorized as unsupervised or supervised, and both require having a training dataset, i.e., a collection of examples, or features that have been quantitatively measured from some object, or event, or signal processing of a particular system that needs classification. Those examples are also called data points, or in general as a knowledge dataset. Supervised learning can be used when the dataset contains features associated with a label or a target. For example, an instrumentation power system may have an acquisition of raw values of voltages and currents that may have distortion and harmonics. Such a raw dataset might go after data acquisition through preprocessing, to fix missing data, cleaning, and interpolation, sometimes also pre-calculating features for feeding an NN. Those features could be Fourier coefficients, average, effective, peak values, as well as inserting labels for some steady-state conditions, such as power factor and displacement factor, or maybe linguistic terms such as overheating conditions, weather stress, and storage at maximum or minimum levels. A supervised learning algorithm would work as a teacher or an instructor, showing the ML model, or eventually an NN, how input features data would be classified in terms of numerical power factor and displacement factor, or into qualitative descriptions such as "low power factor," "highly distorted," "imbalanced conditions," "overloading," "economic cost-effective threshold," and so on.

When a dataset has no label, nor target, neither an instructor, nor a teacher to coordinate the data, then unsupervised learning algorithms must be used. The algorithm might learn the probability density function or may cluster into an n-dimensional center of gravity. Probably the dataset signal might be denoised or the data will find associations with multiple domains. For example, a dataset of ripe avocados could be associated with heat wave weather patterns, or maybe with the cost of Mexican food in restaurants. Such an association could be evaluated on a period of 365 consecutive days, i.e., an epoch of a whole year and seasonal influences. Therefore, unsupervised learning can be extremely powerful if engineering and data science analysis make validation of such results.

Deep learning models will typically execute classification tasks directly from images, text, sound, or features calculated using big data analytics. Deep learning is usually implemented using an NN topology with many layers of neurons (processing units or simply units)—the more the layers, the "deeper" the network. Traditional NNs contain only two or three hidden layers, while deep networks may have hundreds. The neurons in these layers are interconnected, with each hidden layer using the outputs of the previous layers as its inputs.

9.6 Classification building blocks: Instar and Outstar

It is possible to start understanding the interaction between these neurons by taking a close look at the behavior of single neurons. An *Instar* is neuron type with many input

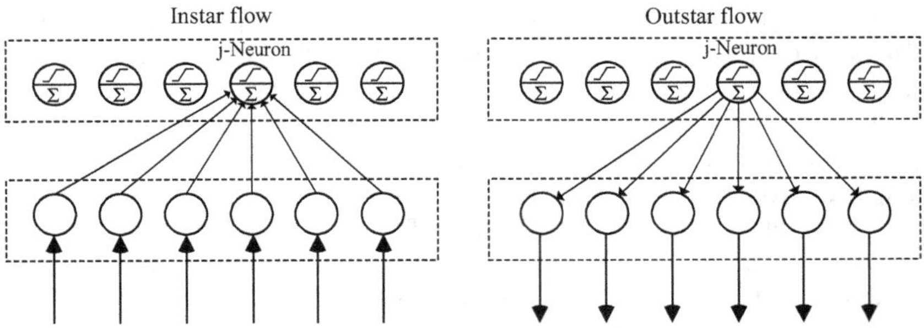

Figure 9.6 Instar neuron versus an outstar neuron connection

connections, a single output, and a particular equation that governs its dynamic behavior. In the instar, the weights afferent to the postsynaptic cell learn the input pattern at the presynaptic sites when the postsynaptic cell is active, as Figure 9.6 shows.

It can learn its inputs by dynamically rotating its weight vector toward the desired input. The *outstar* is the complimentary neuron type of the instar, with a single input connection and many outputs. The weights that project away from the presynaptic cell learn the pattern at the postsynaptic sites when the presynaptic cell is active, as indicated by the outstar connection. It is trained to respond with a specific output when stimulated at the input.

9.7 Classification principles with convolutional neural networks

The principles of NNs can be developed by initially understanding instar and outstar neurons. Those simple elements can be used to establish feature detection in a simple problem and expanded to more complex problems. Figure 9.7 shows an example of a classifier for a fruit image processing where it is assumed that a picture of a fruit with some measurements are available as data. The measurements could be the axis of an ellipse, mass, volume, color, rugosity of the surface, and so on. Therefore, an automated system may detect oranges with their features, such as shape, texture, weight.

In order to conduct the training of this system it is possible to provide an extra input signal, generated by an expert who pre-identified the fruit as an orange or not. Therefore, input data are used for the adaptation of the neural system, a single neuron in this particular case. Elements of input would be constrained to $+1$ (orange visually detected by the expert) or -1 (orange not detected), and a vector of features could have analog values from 0 to 1. Those analog values could also be implemented as a fuzzy inference engine, or it would be just real value numeric variables. If only three features are given as input, the overall dataset could be normalized by a factor of $1/\sqrt{3}$ in order to keep a regular Euclidean distance normalized to the maximum.

For such an example of an instar, several training methods could be used. The instar shares the same structure as the ADALINE (adaptive linear element), which was designed by Widrow and Hoff in 1960 for application in switching circuits of

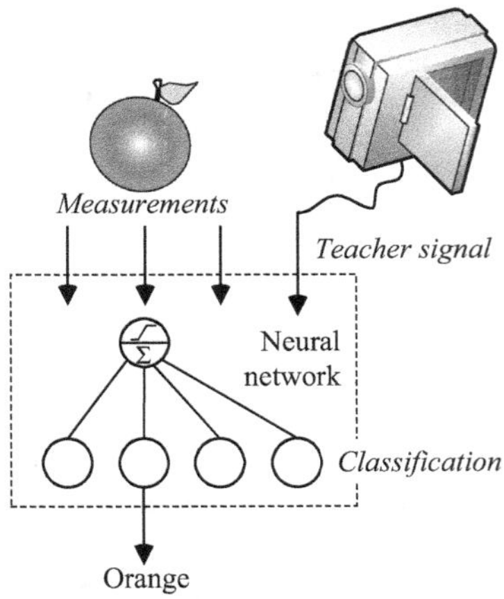

Figure 9.7 Example of a classifier for a fruit image processing

telephone networks. This is considered to be the first industrial application that effectively involved ANNs. Despite being a simple network, the ADALINE promoted some essential advancements to the ANN area.

An instar can be trained by associative learning. In such auto-associative NNs, the output is trained to be identical to the input, usually designed as feedforward, fully connected, multilayer perceptrons, and trained using backpropagation. Such an NN is distinct from an associative memory, typically used for pattern and image recognition (Tsoukalas and Uhrig, 1997). Figure 9.8 shows how to teach a neuron for the identification of an orange, with an instar for such classification. The normalized weight vector moves toward the input vector along a line between the old weight vector and the input vector, like in Hebbian learning. The distance the weight vector moves depends on the value of the error, which must be normalized as well. When the new weight vector is equal to the old weight vector there is no movement; when the new weight vector is equal to the input vector, there is maximum movement, and a proportional change happens when the new weight vector is in between the old weight vector and the input vector. For such a training method, it is important to have the input vectors normalized initially and again after learning. Such a procedure minimizes forgetting, produces normalized weight vectors, and the training process eventually converges.

Instead of associative learning, it is possible to train an instar as an ADALINE, using the delta rule, also known as least mean squares learning algorithm. This algorithm was introduced when the ADALINE was initially proposed, which is considered as the precursor of the generalized delta rule (GDR) used on the training of the multiple-layer perceptron. The GDR is the basis for all modern feedforward NNs training algorithms. Figure 9.8 shows the following inputs: (i) a signal indicating whether a fruit has been visually identified as an orange (unconditioned stimulus), and a (ii) combination (normalized vector array) of the three measurements taken of the fruit (conditioned stimulus).

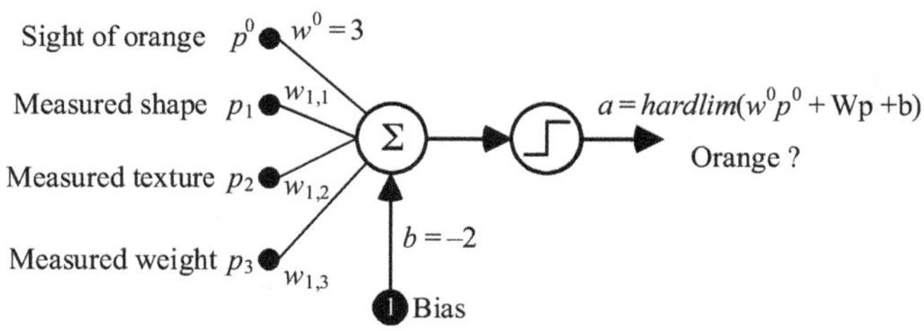

Figure 9.8 Instar detection of an orange

Now we can reverse this problem, and check if, given the features or a fruit, we can detect whether or not that is an orange. The same data could be used, but in this case an outstar neuron could be implemented, as displayed in Figure 9.9. Each signal would feed several outputs; in each output a symmetric saturating linear layer could be activated for a recalled shape, or texture, or weight, as depicted in Figure 9.10. The outstar rule has complimentary operation when compared to the instar rule, because an instar neuron would perform pattern recognition by associating a particular vector stimulus with a response; on the other hand, an outstar neuron would have a scalar input and a vector output to perform a pattern recall by associating a stimulus with a vector response. The outstar could be trained using Hebbian reinforcement learning (RL), which makes the weight decaying proportional to the output of the network. For the instar rule, a forgetting factor should be implemented, by making the weight decay according to the Hebb rule, proportionally to the output of the network. On the other hand, for the outstar rule, the weight decay will be proportional to the input of the network, as indicated in (9.7). If the decay rate γ is set equal to the learning rate α, and collecting terms we get (9.8) showing that the outstar learning occurs whenever the input vector of the network is nonzero, making the weight vector move toward the output vector.

$$w_{ij}(q) = w_{ij}(q-1) + \alpha a_i(q)p_j(q) - \gamma p_j(q)w_{ij}(q-1) \quad (9.7)$$

$$w_{ij}(q) = w_{ij}(q-1) + \alpha\left[a_i(q) - w_{ij}(q-1)\right]p_j(q) \quad (9.8)$$

The principles of instar and outstar demonstrated that a low-complexity classification system could be composed by feature extraction (instar layer) and classification (outstar layer). Smart-grid and power systems will produce huge multidimensional datasets. Therefore, an internal layer for data compression would be important. Assuming that extracted and reduced features can be reverted to the original data (like a linear transform such as Fourier transform), one can quantify and visualize the contribution of individual features toward the original data. Reduced features and reversibility make compressed data useful for classification, and such a reduced feature set would pass through a classification module. Learning in NNs is an optimization problem: the learning algorithm searches for the best possible configuration of weight values, so the network error (or loss function) decays to 0 (and predicts perfectly). As with all other forms of

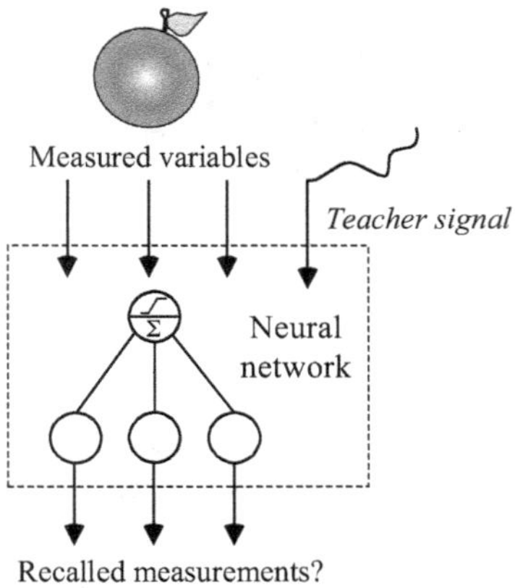

Figure 9.9 Outstar detection of an orange

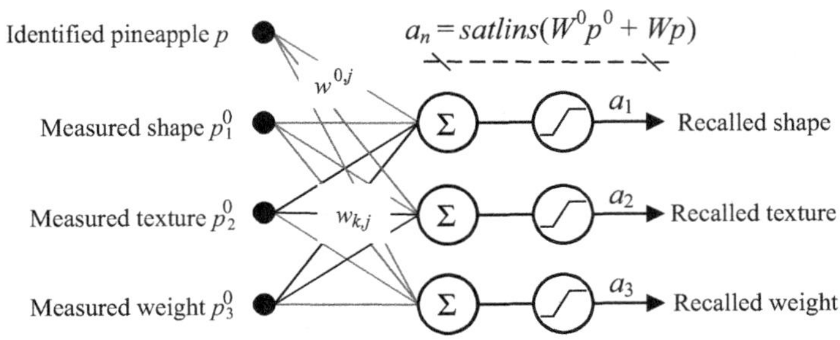

Figure 9.10 Outstar recalling pineapple features and discriminating a pineapple from an orange pattern

optimization, it may not find exactly what it is searching, when the error is finite but not zero. It may also take some time, if the error does not converge or even oscillates. Deep NNs are specialized topologies that permit a modified back-propagation training algorithm to change weights in the inner layers even with extremely large multidimensional data.

Deep learning started with feedforward NNs with many hidden layers, but it was not initially successful because the weight adaptation imposed by the back-propagation of the error gradient from the output to the internal layers would exponentially decrease, making the training processes to not converge or get stuck in a low performance state. However, in the first decade of the 2000s, a few efforts made possible its rebirth as well as some rebranding. Deep learning has been successfully implemented in convolutional NNs (CNNs), as depicted in Figure 9.11. CNNs have modules of three types of layers: (i) convolution, (ii) pooling, and (iii) ReLU. The convolution layer passes the input data through a set of convolutional filters, in order to preprocess and detect specific features or patterns. Pooling (usually max-pooling

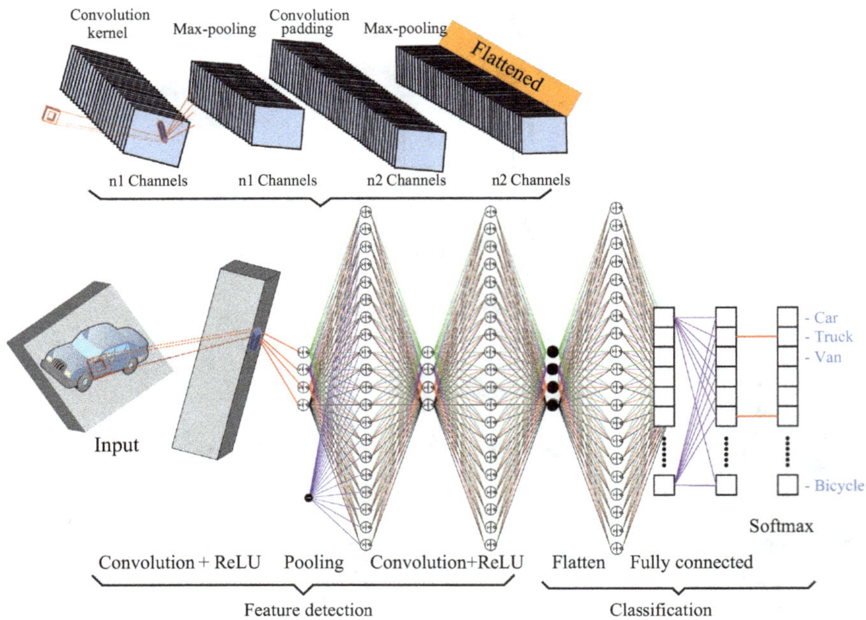

Figure 9.11 Deep convolutional neural network for image recognition and classification

or average pooling) will take the preprocessed data and perform downsampling, in order to reduce the dimensionality of the features being processed. The ReLU is an activation function that has been introduced in CNNs because sigmoid and hyperbolic tangent activation functions cause vanishing gradient problem when many layers are used. In a deep NN topology, the ReLU allows for faster and more effective training by mapping negative values to zero while keeping the positive values. This three-layer functional block is repeated over tens or even hundreds of layers, with each level learning to detect features of increasing degrees of abstraction.

The ReLU activation function was proposed by Nair and Hinton (2010) and has been widely used in deep learning applications. The ReLU is a faster learning activation function, offering better performance and overall generalization power when compared to the *sigmoid* and tanh activation functions (Zeiler *et al.*, 2013; Dahl *et al.*, 2013). The ReLU represents a nearly linear function and therefore preserves the properties of linear models to optimize with gradient-descent methods. The ReLU activation function performs a threshold operation to each input element where values less than zero are set to zero, as indicated by the following equation:

$$f(x) = \max(0, x) = \begin{cases} x_i, & \text{if } x_i \geq 0 \\ 0, & \text{if } x_i < 0 \end{cases} \tag{9.9}$$

The ReLU rectifies the values of the inputs less than zero forcing them to zero, helping one to tackle the vanishing gradient problem observed in the past implementation of deep learning topologies. For computational purposes, the ReLU does not require exponentials and divisions, with enhanced speed for real-time applications. It also introduces sparsity in the hidden units as it squelches the values between zero to maximum. ReLUs may have a limitation: they easily overfit when compared to the sigmoid function, but adding more units in the hidden layer will reduce this problem. ReLUs may also cause gradients to die, leading to dead neurons and causing weight updates to stop propagating backwards. This can be mitigated by allowing the network designer to change hyperparameters, like the number of units in the hidden ReLU layers. The ReLU and its variants have been used in different architectures of deep learning, which include the restricted Boltzmann machines and CNNs, proposed for the first time in 2012 (Krizhevsky *et al.*, 2012). When used with a distinct activation function in the output layer (such as the Softmax), it can be used for object classification, speech recognition, in fault analysis of power systems, and cascading failures in blackouts.

Figure 9.11 shows a block diagram for an image classifier based on a deep CNN (Krizhevsky *et al.*, 2012). There is a huge number of CNNs deployed all over the world for video surveillance and face recognition. Suppose that a network of traffic cameras feed data to this classifier to identify cars, vans, trucks, scooters, or bicycles. A convolution layer is a two-layer feedforward NN that executes many convolution operations that map lower level local features into several higher level feature maps. In this example, the topology has two subnetworks: a feature extractor, implemented by a convolutional base with two sets of convolutional layers, and a classifier, implemented by a densely connected perceptron layer with Softmax output. A flattening layer, which just stacks the various feature maps from the convolutional base into a single feature vector, is introduced between the two subnetworks to make them compatible.

In a CNN there may be connections between the neurons in the same layer; a weight sharing technique is employed between the neurons in different layers to improve the feedforward and backpropagation processes. ReLUs are usually used in the deep convolutional subnetwork, whereas regular NNs would use sigmoid or tanh functions. A common justification is that the ReLU, unlike sigmoid functions, does not saturate with high inputs. For the system depicted in Figure 9.11 all the weights and biases are trained with the backpropagation algorithm, applying stochastic gradient descent. Backpropagation attempts to minimize the loss function between the outputs and target. For multiclass classification, a Softmax output layer with the cross-entropy loss function (or multinomial logistic loss) is typically used.

The classification subsystem is a fully connected layer that outputs a vector of K dimensions where K is the number of classes that the network will be able to predict. During training, the desired output vectors are one-hot encoded: the dimension associated with the predicted class has its value set to one while all others are set to zero. The classification output is based on a Softmax function. After training, the network outputs a vector that contains the probabilities for each class of any image being classified, with the predicted class having the highest probability. Equation (9.10) shows the calculation of a Softmax. The Softmax function is also an activation function to compute a probability distribution from a vector of real numbers, and that is why it is used for multiclass classification problems. The main characteristic for choosing Softmax instead of Sigmoid is that Sigmoid is used for binary classification, while the Softmax is used for multiclass classification tasks. The Softmax activation function is a smooth version of a competitive (1 to *N*) function, which can be used with a loss function derived from a measure of relative entropy between the target and the actual output, assuming the output is scaled between 0 and 1. Standard backpropagation can be used with the Softmax activation function, because the chain rule will make the derivative equal to desired target minus the actual output. Therefore, a CNN will have at its output a logistic regression over the extracted feature representations from prior layers (the convolutional base), because the Softmax will encode a probability distribution instead of using feature templates.

$$f(I_k) = \frac{e^{I_k}}{\sum_{l=1}^{k} (e^{I_l})} \tag{9.10}$$

The use of a deep CNN, as shown in Figure 9.11, for image recognition and classification, establishes some principles for the application of deep NNs, with a variety of topologies, paradigms, and classification algorithms. In summary, it is necessary the (i) input subnetwork executing feature extraction from raw inputs (the convolutional base in the case of CNNs) and the (ii) output constructing probability density functions for multiple class detection. A topology based on instars and outstars with inner layers for computation of features, and outputs for selecting the winning classes based on a Softmax squashing function, or maybe Boolean decision making, would allow a sophisticated deep learning NN. Feature extraction layers can be fully trained on three types of operations on the data: convolution, pooling, or ReLU; therefore, those data processing functions can be implemented in other similar ways.

9.8 Principles of recurrent neural networks

NNs, SVM, logistic regression, and fuzzy logic reasoning have been applied successfully as classification tools in many tasks without considering the time, dimension, or assuming time independency. If past input or output is assumed to influence the current state, the time dynamic could be captured by using a sliding time window, gathering together previous, current, and internal hidden states as a single input vector representation. However, such an approach fails to capture long-range time dependencies, effects outside the window size. A conversation or a dialogue made today with a historical background of about 3 years ago, may not include facts of a previous childhood social effect. A conversation with two people who know each other for the past three years may not have the data time window, neither memory or buffer, to bookkeep learning outcomes, unless the dialogue is very intense with remarks typical of human-like responses.

Suppose someone is interested in predicting the general energy prices of two competing electric power companies for the next week in order to decide on a contract and electrical power scheduling. Let us assume we have data available for both companies for the last week, and for a week recorded from last year's transactions. We can also assume that we have current data recorded in the past 7 days, but usually data are diverse as records are taken by different entities. What would be the decision made just on the estimated average price for the following week, or maybe a day-by-day prediction for the next week, based on past history (last year) and recent history (last 7 days)? Suppose the past history for one energy company has data recorded every hour for the past 10 days, while the other energy company has data recorded every 15 min for the past 7 days, as indicated in Figure 9.12.

Such inputs are different in size, as well as in sampling frequency. Their visual appearances could be similar, but an epoch of data from one company would be different from the other. In this hypothetical study, we are assuming only two companies, but in a more realistic scenario, we could have hundreds of companies, or consumers, or even other data sources, such as smart-meters, and hybrid electric vehicle storage. We may have very large variability in data volumes, sampling rates, variety, and values.

One way to address the variable-sized input problem is to use a deep CNN with an input layer that is large enough to cover all sequential data, followed by enough sequences of convolutional sets of layers (convolutional, max pooling, and ReLU arrangements), so that the down-sampled parameters that the network must learn could be consistent with all data cases. However, still an issue that may happen with time series, or systems that have differential equation dynamics, difference equations, or memory, is that the model should be invariant to translation. This is an undesirable property for an energy price prediction model, because we want to make use of the precise temporal information (the energy price today is likely to be more influenced by the energy price yesterday than 1 year ago).

A proper way to deal with variable-sized inputs, and with dynamics and memory, is to use a recurrent NN (RNN), originally presented by Elman (1990). It was developed during the 1990s and an adaptation of the backpropagation training algorithm called backpropagation through time (BPTT) was used to train an Elman network, which became known as an RNN. For more than 20 years, this was the dominant approach, until such a topology evolved to the long short-term memory (LSTM) network, described also in this chapter.

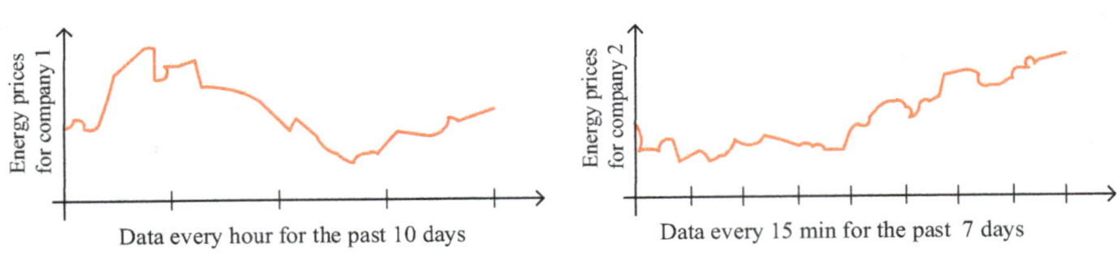

Figure 9.12 Energy price for two different electric power companies in the recent few days

A typical RNN is able to process a sequence of arbitrary length by recursively applying a transition function to its internal hidden state vector of the input sequence. The activation of the hidden state at each timestep is computed as a function of the current input and the previous state. An important contribution of RNNs is in the area of language modeling, an important task in natural language processing. The goal of language modeling is to predict the next word, given a sequence of previous words in a text document. For example: (i) "I go to . . . ," (ii) "I play soccer with . . . ," (iii) "The plane flew within 40 feet of . . . ," and so on.

It is possible to convert each word in the sequence into a vector, by constructing a dictionary of all possible (or at least the N most frequent) words in the text document corpus. Every word is then associated with an entry index in the dictionary (normally an integer number). A vector of dimension N (the size of the dictionary) can be defined now, with zeros in all dimensions except for that particular dimension that corresponds to the word index, which receives the value of 1. This is called the one-hot vector encoding of the word according to that dictionary.

RNNs have been applied with success in the task of sentiment analysis of movie reviews: rating the text review in a scale of five classes (negative, somewhat negative, neutral, somewhat positive, and positive), or binary classes (recommended, not recommended). RNNs were also applied with some success in speech recognition, automatic computer language translation, automatic image, and video captioning.

BPTT (Werbos, 1990) learning for RNNs is usually difficult to implement due to the vanishing/exploding gradient problems, i.e., the magnitude of the gradient becomes extremely small or extremely large toward the first or the last timesteps in the network. A practical maximum timestep window is about ten epochs for BPTT implementation. The main reason for such vanishing/exploding gradient problem is the use of sigmoidal activation functions in the original recurrent networks. As the error derivatives are propagated backwards, they have to be multiplied by the derivative of the sigmoid (or tanh) function, which can saturate quickly. The use of the ReLU activation function instead of the sigmoid or tanh allows for better gradient flow, thus tackling the vanishing gradient problem. However, another problem arises when ReLUs are used: in the backward error computation, depending on the initialization, hidden values can explode or vanish depending on whether the eigenvalues of the internal matrix are bigger or smaller than 1.

The problem of deep learning makes the training of RNN challenging, especially when there are a lot of long-term dependencies, meaning that the output prediction is influenced by pieces of information from a distant past, or that there is a very long effect of current information in the future events. Some natural human languages have such effect, and analyses of historical humankind facts are an example of that. On the other hand, in most of dynamic physical or engineering systems, the delayed input is usually associated with the order of the differential equations governing the phenomena, so for every day simple systems, it may not be necessarily a large time window. For example, an induction machine has a fifth-order nonlinear system of differential equations: it would be an ad-hoc supposition that more than five previous inputs to the system would have to be recorded and memorized; but because the system is nonlinear, it would be advisable to record at least 10 times such order, about the 50 last inputs plus the last 50 hidden states. But, if a complex power system is composed of scores of hundreds of induction machines, with energy and power constraints, and non-linear loads, a regular 10 timestep RNN may not be able to capture such deep, intricate behavior. The size of an RNN can easily become too large for some engineering systems, which motivates the design of new types of RNNs, particularly the LSTM architecture, based on the combination of distinct individual RNNs.

9.8.1 Backpropagation through time-based recurrent neural networks

Figure 9.13 shows a six-neuron single-layer attractor network with feedback, and it accepts three input signals and generates three output signals; such a configuration is a typical Hopfield network. It is an example of an early proposed recurrent network, composed of a single-layer attractor network feeding output neurons in a feedback competition learning system. In 1997 the research conducted by Hochreiter and Schmidhuber (1997) proposed that LSTM could solve the problem of the vanishing gradient in deep learning topologies. In an NN, long-term memory is encoded through adjusting the weights of the network and once trained, these weights should not change. Short-term memory is encoded through the activations that flow through the network and they decay quickly; therefore, LSTM units are designed to enable the short-term memory (the activations) in the network to be propagated over long periods of time, or sequence of inputs.

Figure 9.14 shows a recurrent backpropagation network. The network has three inputs and three outputs, the middle layer has two neurons. Three extra units are recurrent neurons feeding the output; and two recurrent neurons are feeding the middle layer; all those recurrent neurons get their input from the output of the NN system, they are marked with R. In some topologies, recurrent neurons are schematized as an additional middle layer in addition to the regular middle layer. It is

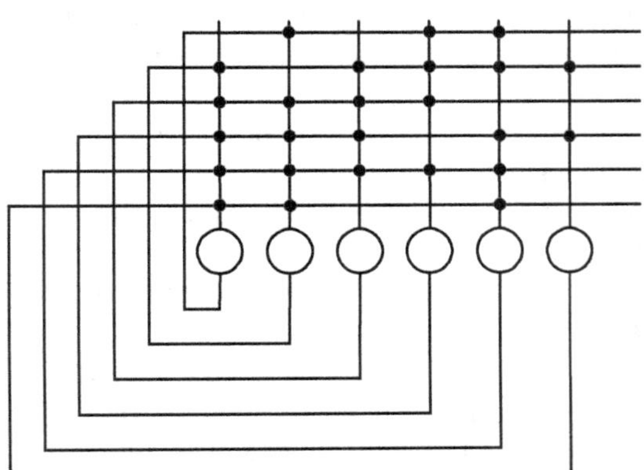

Figure 9.13 Simple attractor six-neuron recurrent network

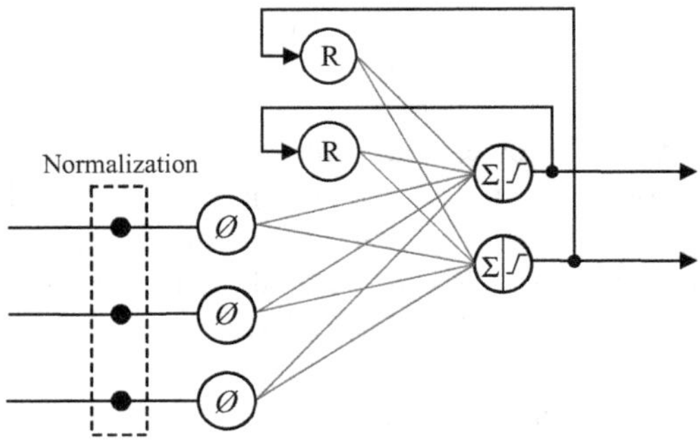

Figure 9.14 Simple recurrent backpropagation neural network

possible to have only recurrent neurons feeding either the output layer alone, or the hidden layer alone. We can observe that a recurrent network has substantially more connections than a regular feedforward backpropagation NN and certainly requires more memory in the algorithmic implementation. For example: a regular densely connected feedforward NN with a 3–2–3 (input–middle–output) topology has 12 adaptive weights (3×2 from input to middle +2×3 from middle to output); a recurrent 3–2–3 topology will have 25 adaptive weights, i.e., the same 12 as before plus 4 weights between the recurrent neurons added to the input layer and middle layer, and 9 more weights between the recurrent neurons added to the middle layer and the output layer. A possible simplified implementation is to make each weight on the connections leading into the recurrent neurons to be fixed at 1.0, so there are less adapting weight calculations.

In order to train such an NN, it is necessary to keep track of the activity levels for each of the recurrent neurons. If a pattern sequence is considered to have an epoch of N steps, then N copies of each neuron activity must be maintained. It is possible to imagine a movie film, each frame of action throughout the sequence, where the recurrent neurons provide a temporal context for each frame. In a recurrent neuron network, the activity of the middle and output layers propagates through time to the next timestep, eventually influencing the network output at later time.

Suppose at (timestep=1), an initial input pattern is applied to the network. The output and middle layers are assumed to have some initial standard output from initial weights. The input pattern is processed by the network, and a corresponding output pattern is generated at this tick of the clock. This output pattern is stored for future computation of errors and weight changes. At (timestep=2), the second input pattern in the sequence is presented and propagated through the network. However, at this tick of the clock, the middle and output layers have additional inputs from the activity in those layers from the previous timestep=1. Such a leftover activity, combined with the input stimulus, generates the network output pattern for timestep=2, and such a result is also stored for later computation of the error. Such a cycle continues, on and on, throughout all the N timesteps of the pattern sequence, where at each tick of the clock, an output is generated, in which the activity is stored for later use. When all N steps are completed, the error will be computed, the weight changes generated, and the final weight change for such sequence is the sum of those N weight changes. This training procedure is called batch training in the backpropagation terminology.

However, the dynamic equations that govern recurrent networks are more complex than regular backpropagation, because each timestep activity is dependent on the current input pattern plus the activity level of the previous timestep, and error computation is dependent on the current and the previous activity of the previous tick of the clock. Therefore, the controlling equations are recursive, i.e., for the RNN of Figure 9.14 the middle layer neurons receive the following stimulus, as indicated in the following equation:

$$I(t)_j^{mid} = \sum_{i=1}^{3} w_{ij}^{in-mid} y(t)_i^{in} + \sum_{k=1}^{2} w_{kj}^{mid-mid} y(t-1)_k^{mid} \tag{9.11}$$

where the net input to each middle layer neuron is the sum of the net input from the input pattern applied at the current timestep (t) and the input from the middle layer's own feedback from the time one tick ago ($t-1$); the first term in the expression is the same as a nonrecurrent feedforward backpropagation NN, while the second term reflects the middle layer activity feeding back to itself (observe the diagram in Figure 9.14), note that if $t=1$, the $y(t-1)$ term in the right expression is the initial *reset activity* value (or the activity from the previous cycle). The net input for the output layer is computed similarly with appropriate changes to reflect the sources of the signals and different sizes of the layers, as indicated in (9.12), again if $t=1$, the $y(t-1)$ term in the right expression is the initial reset activity. Figure 9.15 depicts the timestep evolution from $N=1$ to $N=3$ for the RNN as indicated in Figure 9.14.

$$I(t)_j^{out} = \sum_{i=1}^{2} w_{ij}^{mid-out} y(t)_i^{mid} + \sum_{k=1}^{3} w_{kj}^{out-out} y(t-1)_k^{out} \tag{9.12}$$

In both layers, the output activity is computed using a nonlinear activation function over the net input. A typical shallow NN would use a sigmoid function, while a deep NNs topology would use ReLU or Softmax functions. In (9.13) the function $f\left(I(t)_j\right)$ is a logistic (sigmoid), but it could be another appropriate activation function used in the implementation. Once the activation flow through the network is computed for all timesteps in the sequence, the error signals will be calculated:

$$y(t)_j = f\left(I(t)_j\right), \text{ where } f(I) = \frac{1}{1+e^{-I}} \tag{9.13}$$

Figure 9.15 shows how a recurrent network evolves across timesteps, in each one the network output is generated, the activity of the middle and output layer neurons propagates through time to the next timestep, where it influences the

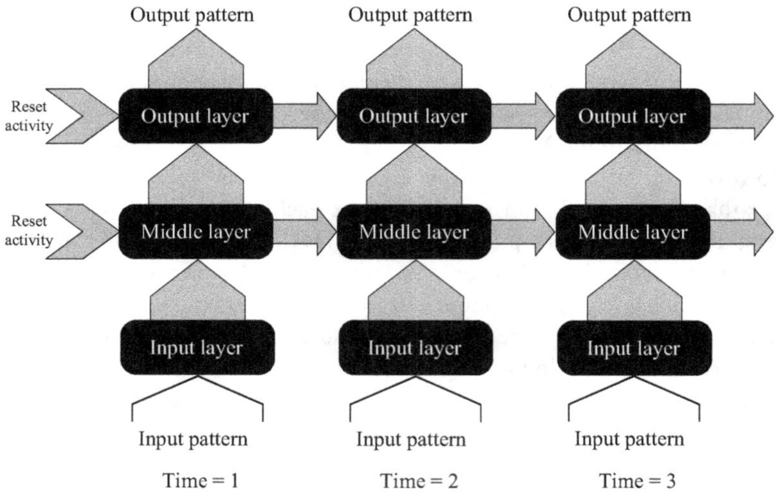

Figure 9.15 Recurrent network evolution across timesteps

network output at later time, while Figure 9.16 summarizes the whole implementation. For a numerical computation implementation, executed at the beginning of timestep=N, i.e., at the stopping time for the sequence (N is a batch called epoch), calculation works backward to the initial (timestep=1). In such implementation, the recurrent network backpropagates not only through the physical layers of the network, but also through time itself. Such implementation requires a great deal of programmatic bookkeeping to track which weight or error or activity is required at each phase, slowing down the whole training process. It takes a large number of passes to train an RNN model with a large dataset. There is a high computational complexity calculation, complex relationships and dynamics can be learned. Figure 9.16 is a general block diagram step-by-step implementation.

9.8.2 Long short-term memory (LSTM)-based recurrent neural networks

RNNs are a superset of feedforward NNs. Recurrent connections are added to the output layer, or the hidden layer, or both. They are recursive in keeping data for spanning adjacent timesteps (e.g., previous timestep, or even more), allowing the modeling of the time and the dimension in an NN framework. The typical RNN uses BPTT for training, but the vanishing gradient problem makes it difficult to model long-range time dependencies (ten timesteps or more). Training a deep network for recurrent applications can be approached by using both the gradient descent algorithm (which defines the actual updates of the weights) and the backpropagation algorithm (which sorts the credit or blame of weight assignments) in tandem. Backpropagating error through t timesteps in the past involves multiplying the error gradient by the same set of weights t times, which is roughly equivalent to

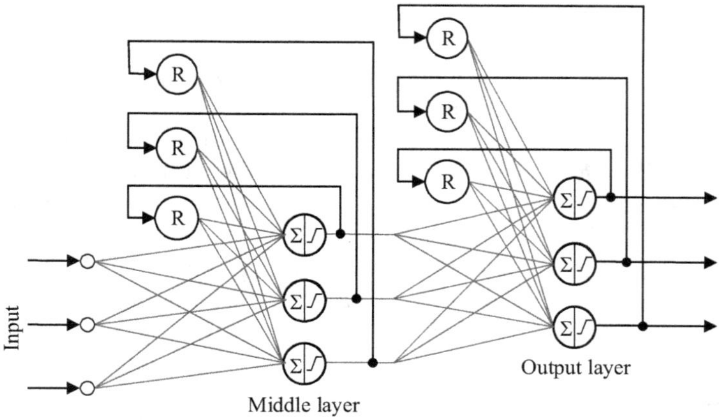

Number of outputs = L
Number of hidden layer neurons = H
Number of inputs = M
Time-steps window from 1 to N

- Forward activity processing to each layer from $t = 1$ to $t = \mathrm{N}$

$$I(t)_j^{mid} = \sum_{i=1}^{M} w_{ij}^{in-mid} y(t)_i^{in} + \sum_{k=1}^{H} w_{kj}^{mid-mid} y(t-1)_k^{mid}$$

$$I(t)_j^{out} = \sum_{i=1}^{H} w_{ij}^{mid-out} y(t)_i^{mid} + \sum_{k=1}^{L} w_{kj}^{out-out} y(t-1)_k^{out}$$

- Backward error propagation for each layer, from $t = N$ to $t = 1$, starting with error at $t = N$

$$E(N)_j^{out} = (y(N)_j^{desired} - (y(N)_j^{actual}) \frac{df(I(N))}{dI}$$

$$E(N)_j^{mid} = \left(\sum_{i=1}^{L} w_{ij}^{mid-out} E(N)_i^{out} \right) \frac{df(I(N)_j^{mid})}{dI}$$

- Backward error propagation for all times prior to $t = N$

$$E(t)_j^{out} = \left(\varepsilon(t)_j + \sum_{i=1}^{L} w_{ij}^{out-out} E(t+1)_i^{out} \right) \frac{df(I(t)_j)}{dI}$$

where

$$\varepsilon(N)_j^{out} = (y(N)_j^{desired} - (y(N)_j^{actual})$$

$$E(t)_j^{mid} = \left(\sum_{i=1}^{L} w_{ij}^{mid-out} E(t)_i^{out} + \sum_{k=1}^{t} w_{kj}^{mid-mid} E(t+1)_k^{mid} \right) \frac{df(I(t)_j^{mid})}{dI}$$

- Weight change law for nonrecurrent weights

$$\Delta\omega(t)_{ij}^{in-mid \text{ also } mid-out} = \beta E(t)_i y(t)_j$$

- Weight change law for recurrent weights

$$\Delta\omega(t)_{ij}^{mid-mid \text{ also } out-out} = \beta E(t)_i y(t-1)_j$$

Figure 9.16 Block diagram and equations implementation for an RNN with M inputs, L outputs, H hidden-layers, plus two sets of recurrent neurons for the middle and output layers

multiplying each error gradient by a weight raised to the power of t. Since these weights are usually in numbers much smaller than 1, the backpropagated error component diminishes in an exponential fashion as the length of the time sequence increases, eventually vanishing when the temporal sequence is large enough, resulting in a no learning effect. A recurrent network will have the output layer receiving signals from the middle layer as well as from the recurrent ones, using regular activation functions, as indicated in Figure 9.17. There are several variants of RNNs, but if a model based on Figure 9.17 is adopted, where recurrent neurons are only connected to the output layer without a hidden layer, the combined data of input layer with the output of recurrent nodes define a "data concatenation." The bottom of the figure shows a unifilar vector flow diagram of the neuronal implementation at the top.

If we assume that an enhanced unit will be used instead of those simplified extra recurrent neurons, capable of further memory of past events, an especially promising topology is introduced, called LSTM network. LSTMs are very powerful in the implementation of deep recursive neural structures, such as the one depicted in Figure 9.18. LSTMs were introduced by Hochreiter and Schmidhuber (1997).

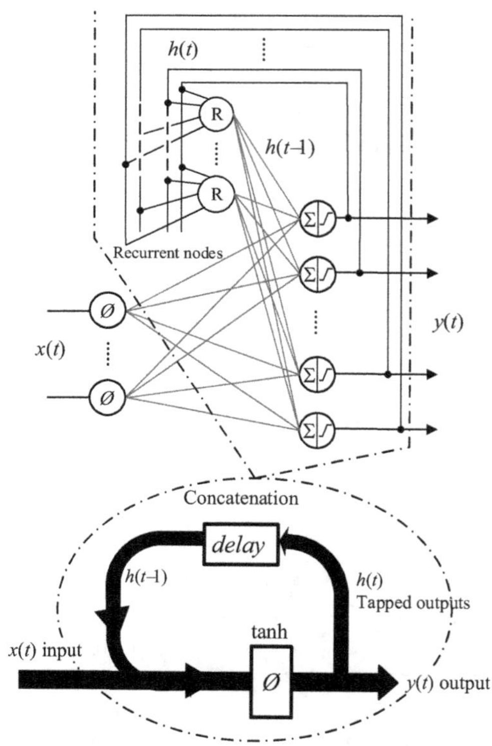

Figure 9.17 A recurrent neural network where the combined data of input layer with the output of recurrent nodes define a "data concatenation." The bottom of the figure shows a unifilar vector flow diagram of the neuronal implementation at the top

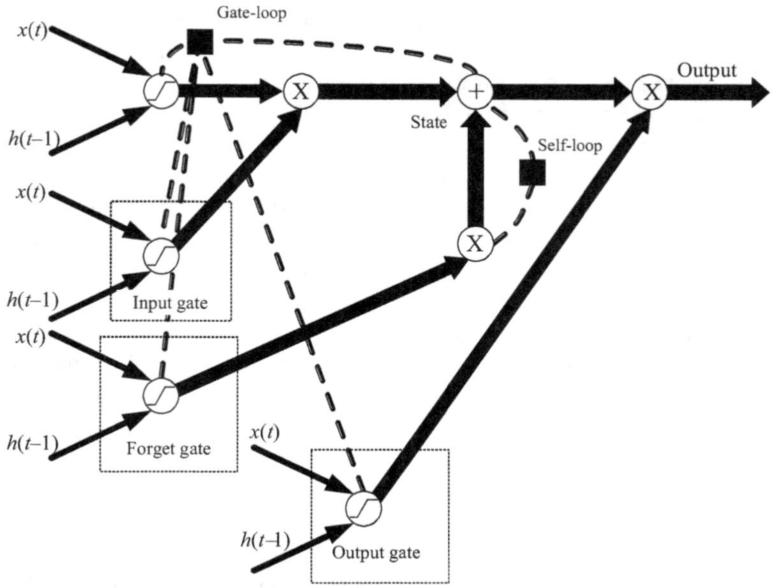

Figure 9.18 Block diagram of an LSTM recurrent network unit

Short-term memory is encoded in the hidden state (or activations) of the recurrent units that flow through the network (similar as the original RNNs) while long-term memory is encoded in a new structure of processing units called the cell state, which can store and recall the hidden state. Therefore, LSTM units enable the short-term memory (the activations) in the network to be propagated over long periods of time, or sequences of inputs. LSTMs have excellent performance for a large variety of engineering problems, being widely used and improved over time (Graves, 2012; Gers *et al.*, 1999). LSTMs reduce the effect of vanishing gradients by adding a parallel path to the repeated multiplication by the same weight vector during the backpropagation in a regular RNN.

The core cell is indicated in Figure 9.18 in which the activation or hidden state (short-term memory) can be stored or merged into the cell state vector and propagated forward. The contents of the cell state are regulated by three controllers called (i) forget gate, (ii) input gate, and (iii) output gate. In the LSTM the cell state is the horizontal line running through at the top of the diagram. The flow of information is controlled by a multiplication block (unchanged when multiplied by 1) and an addition block (unchanged when added with zero); therefore, information is removed or modified in the cell state by those regulated structures called gates.

Gates are a way to optionally let information through. They are composed out of an activation recurrent neural net layer (sigmoid, tanh), as well as a pointwise multiplication operation. In the original RNN structure, the output layer would receive the $x(t)$ information directly from the middle layer, the recurrent node would access the previous $x(t-1)$ and pass it through an activation function. However, in the

LSTM, the recurrent neuron is a bit more involved in complexity. The forget gate will take the concatenation of the input and hidden layers and pass such vector through a layer of neurons that use a sigmoid function, i.e., σ in Figure 9.18, as a result, the neurons in the forget layer give a value from 0 to 1, multiplying the previous cell state by such a factor. Values near 0 make the previous cell state be forgotten, while values near 1 keep the cell state being propagated. Next, the input gate will decide on which information should be added to the cell state.

First the gate decides which elements in the cell state should be updated and what information should be included in the update. This decision occurs by concatenating the input vector (x_t) and the hidden state vector (h_{t-1}), then going through the sigmoid (σ) units to generate a vector of elements, the same width as the cell state, where each element is in the range from 0 to 1 (sigmoid output). Sigmoid output values near 0 indicate that the corresponding element will not be updated, while values near 1 indicate that the corresponding cell elements will be updated. At the same time, the concatenated input and hidden state are also passed through a layer of tanh activation functions, one tanh unit for each activation in the LSTM cell. The result represents information that may be added to the cell state. Of course tanh units are used because their output ranges from -1 to $+1$, and the cell elements could be either increased or decreased. The final update vector is calculated by multiplying the vector output from the tanh layer by the filtered vector generated from the sigmoid layer, as indicated in Figure 9.18, in the addition to the cell state.

The final state in the LSTM processing is to decide which elements of the cell state should be merged with the output in response to the current input, as depicted in the dashed box marked output in Figure 9.18. A candidate output vector is generated by passing the cell through a tanh layer, at the same time the concatenated input and propagated hidden state vector are passed through a layer of sigmoid units (filtered vector), the actual output vector is then calculated by multiplying the candidate output vector by this filtered vector, being passed to the output layer. Then this output is propagated forward to the next timestep as the new hidden state h_c. An LSTM cell is a network by itself. It can be used in an RNN topology as the recurrent neuron discussed in the beginning of this section. Often in the literature an RNN topology that uses LSTM units are known as LSTM networks as whole, Figure 9.19 shows a comprehensive stacking of LSTM units to form a whole LSTM NN.

9.8.3 Fuzzy parametric CMAC neural network for deep learning

Parametric CMAC (P-CMAC) ANNs were introduced in 2002 (Almeida and Simões, 2002), as a fast and efficient alternative to commonly used MLP networks for applications like function approximation and signal processing. A P-CMAC architecture is based on the CMAC network and on the Takagi–Sugeno–Kang parametric fuzzy systems (TSK-fuzzy systems). CMAC networks were proposed in 1972 by Albus (1975a,b). TSK-fuzzy systems were proposed by Tomohiro Takagi, Michio Sugeno, and Geuntaek Kang (Sugeno and Kang, 1988), where they were

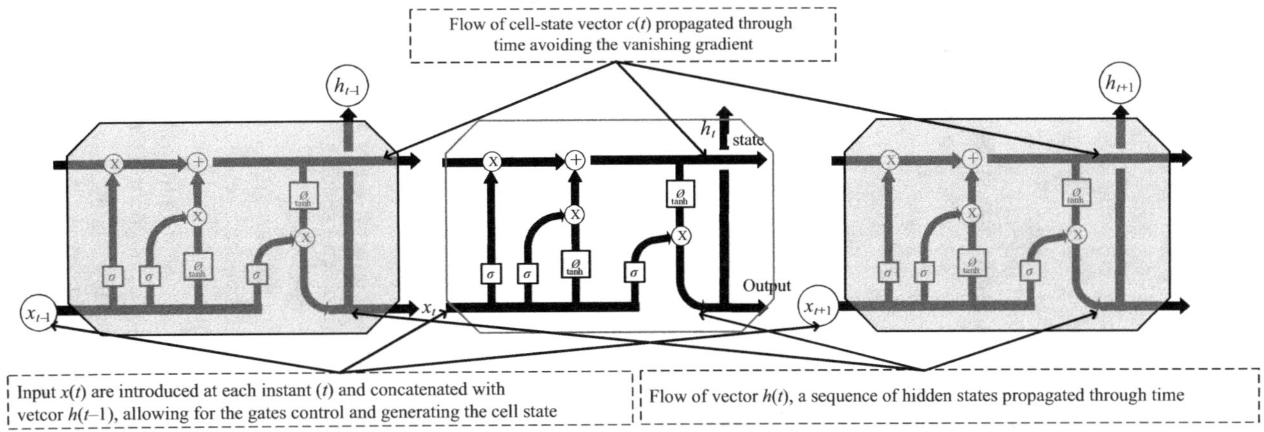

Figure 9.19 LSTM neural network

demonstrated to have very good function approximation features (Sugeno and Kang, 1988; Takagi and Sugeno, 1985). Fuzzy P-CMAC network retains the local data flow properties of CMAC and implements parametric fuzzy equations in a hidden layer inside the network, resembling fuzzy-TSK systems.

The P-CMAC network is improved on the limitations of Albus CMAC algorithm, such as a better non-smooth response for smooth inputs and enhanced representation capabilities for complex systems. The simple binary behavior of CMAC receptive fields was the cutting-edge model when first proposed by Albus, but P-CMAC networks have enhanced internal mapping.

P-CMAC networks have important modification in their internal mapping, when compared to CMAC. A CMAC internal memory was implemented as a bit memory location connected to the binary input activation functions. Fuzzy-CMAC networks use real valued memories, each one corresponding to the membership grade of the input with respect to the fuzzy activation functions. The P-CMAC network retains the local data flow properties of CMAC and implements parametric fuzzy equations in a hidden layer inside the network, resembling fuzzy-TSK systems. Therefore, a fuzzy P-CMAC network has real valued memories, each corresponding to the membership grade of the input with respect to the fuzzy activation functions.

Figure 9.20 shows a schematic diagram of the fuzzy P-CMAC architecture, where the long-term memory is composed of the input layer (which may have some preprocessing) plus a feedforward backpropagation hidden layer, capable of weight adaptation. A layer of parametric fuzzy relations is trained using multivariable least square optimization, in faster real-time mode, making the system also capable of short-term memory learning. Such an integration of parametric fuzzy with the output layer is capable of fast dynamics learning.

James Lo proposed, in a sequence of papers, the implementation of an ANN with LSTM applied to dynamical systems identification, where the drawbacks of adaptive adjustment of ANNs to this task have been discussed (Lo and Bassu, 2001, 2002; Lo, 1996). The adaptive adjustment of all weights in this type of network led to the following conclusions:

- The existing algorithms that perform adaptive adjustment (like back-propagation, Kalman Filters, and others) spend a lot of memory and CPU time and have very low convergence rates.
- The criterion to be minimized, in the case of a whole network with nonlinear neurons, presents quadratic or higher order terms. This implies the existence of poor-quality local minima. Online training does not allow multiple session optimization nor employment of global optimization methods. So, the probability of finding and getting stuck on a poor local minimum is high.
- Online adaptation of all the network weights does not use a big dataset (which can be available) in an intelligent way. As it neither gives priority to identify the system dynamics nor to identify some specific operation points of the plant, the resulting network is not a good approximation of the system to be modeled.

A natural way to avoid these drawbacks is to create an ANN with two different sets of connection weights: those related to LSTM. Long-term connection weights

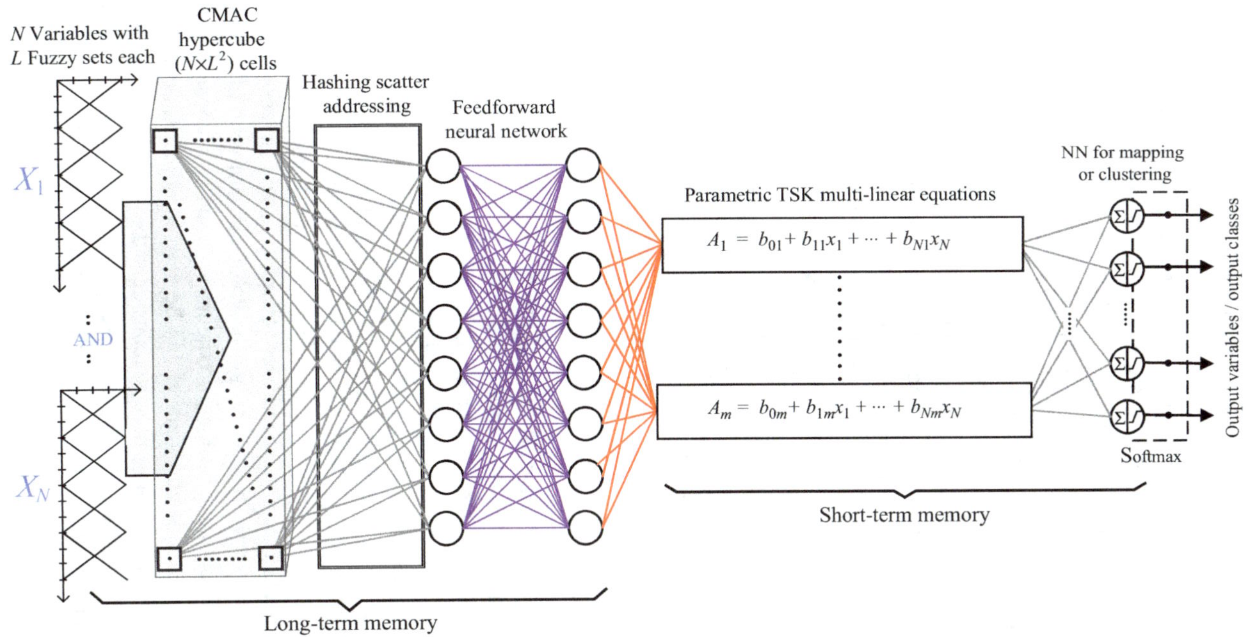

Figure 9.20 Fuzzy parametric-CMAC network for deep learning applications

should be trained offline, with big datasets, which in turn should represent the system to be modeled in a variety of conditions and operating points; short-term weights should be trained online to fine tune the network response to instantaneous operating conditions. In other words, the long-term training is focused on getting the main dynamics of the system being modeled.

After this offline learning phase, the long-term weights or memories must be kept constant, preserving the input information. The dataset used to train the long-term memories in an LSTM NN must include all operating points of interest and a lot of different values of the environmental parameters, such as values of environmental, constructive, and low-frequency noise referring to the system to be modeled. For each one of these conditions, the training dataset should include a subset of input–output signal pairs. Input values should be uniformly chosen inside the region of interest or in a slightly bigger region, to allow a good approximation of the system behavior close to its boundaries. It is important that the long-term dataset comprises input–output pairs covering the whole region of operation of the plant. The long-term training is independent of the various parameters, and the offline training of an LSTM network can be based on a long-term optimization criterion such as the following:

$$J_{LP}(w_{LP}, w_{CP}(\theta)) = \frac{1}{s \times n} \sum_{i=1}^{s} \sum_{j=1}^{n} (U_{ij})^2 \tag{9.14}$$

where

$$U_{ij} = y_{ij} - \widehat{f}(u_{ij}, w_{LP}, w_{CP}(\theta)) \tag{9.15}$$

The previous two equations describe a utility function that measures the error between the desired and the actual output of the modeled system. During long-term training, the parameter values should be adjusted to minimize the utility function. This is a soft-constrained optimization problem, for which a lot of traditional methods can be applied and which can often achieve exact solutions. On the other hand, weights related to short-term memory should be trained online, using a utility function that can capture specific environmental/constructive parameters currently active in the plant or control system. Adjustment of these weights will improve the controller performance. For this specialized training, one can assume that the parameter values may vary in time in an unknown or random way, although there are a number of input/output data measurements between two consecutive parameter value changes. Then, such a dataset can be used to adjust short-term memories of the network. The learning process considers the active changes and, just like in the long-term training process, the specialized training should use a performance criterion based on a utility function as in the following:

$$J_{CP}(w_{CP}) = \frac{1}{m} \sum_{i=1}^{m} \lambda^{m-1} \times U_i^2 \tag{9.16}$$

where

$$U_{ij} = y_{ij} - \hat{f}(u_{ij}, w^*, w_{CP}) \tag{9.17}$$

where w^* are long-term memory values obtained during general training, and λ is a forgetting factor that weighs each instantaneous utility function based on its temporal proximity. This implies that older data will contribute less in the minimization of the J_{CP} function. If an ANN has only linear activation neurons at its output layer, and also if the weights of this output layer are chosen to be the short-term memories of this network, then short-term training will be equivalent to a multiple linear regression algorithm. In these conditions, there will be no local minima and training will converge to the global minimum of the short-term utility function.

Therefore, an ANN with LSTM has a long-lasting or long-term memory, reflecting basic features of the plant being modeled and ensuring that the model generated by further training other parts of the network will not deviate much from the original behavior captured during long-term training. These same ANNs have a short-term memory with continuous or frequent adjustments occurring when environmental changes are detected, reflecting the response of the network to these changes, and making the ANN to adapt to the newly verified operation conditions.

James Lo demonstrated that this approach leads to networks that exhibit universal approximation properties, just like the classical MLP networks. Empirical results showed that an MLP network with LSTM performs better than a regular MLP network with the same number of neurons, yet avoiding training process issues, particularly the vanishing gradient problem. Considering that fuzzy P-CMAC networks are feedforward networks with similar features to those of MLP networks, long- and short-term memories are also intrinsic to the fuzzy P-CMAC. This technique has been applied in the modeling of the dynamic behavior of a controlled fuel cell in Almeida and Simões (2002, 2005), Meireles *et al.* (2003).

The principles discussed in this chapter can be applied to power systems, power electronics, power quality, and renewable energy systems as well. This chapter discussed the most promising and successful deep learning techniques by the time of the writing of this book, although many other NN topologies can be used. Deep NNs have been successfully implemented either as CNN or as RNNs, particularly as LSTM implementations. CNNs operate mostly in pattern recognition and mapping tasks, very useful for image processing, while RNNs are very useful in tasks involving encoding, classification, and regression on sequences of arbitrary length, including time series. In the last few years, the development of the LSTM architecture made possible recurrent networks to be implemented for very long sequences and extremely large datasets. LSTM cells (or processing elements) incorporate an embedded state memory controlled by RNN information gates. This allows cell state to be stored and carried over long time spans, minimizing the effect of the vanishing gradient problem in long sequences.

Training a deep learning model may take minutes, hours, days, or weeks, depending on the size of the training dataset, and the processing power available. Selecting a computational resource is a critical consideration for the workflow.

Fuzzy P-CMAC NNs also allow deep learning with embedded long- and short-term memories and fuzzy logic hybridization.

Currently, there are quite a few computational platform options for the development of deep learning models: CPU-based, graphics processing units (GPU)-based, and Cloud-based. CPU-based computation is the simplest and most readily available option, which can be done in any regular workstation or laptop computer. Using GPU reduces network training time by a large factor because of the specialized hardware, data structures, instruction sets, and extensive use of parallel computing operations. It is possible to incorporate GPU support in many software packages (such as MATLAB® and GNU Octave), scripting or compiled languages (such as Julia or Python, with libraries such as TensorFlow and PyTorch) without additional programming, but it is necessary to use compute-capable GPU hardware and library support (such as NVIDIA CUDA). Multiple GPUs can speed up processing. Cloud-based GPU computation means that the designer is using a virtual machine with GPU support made available on the Internet and does not have to buy and set up the hardware and even the software platform (by using precompiled images). These virtual machines can be provided by specialized cloud computing companies or by a shared team of professionals displaced all over the world.

The knowledge of physics and electrical engineering modeling of systems would allow inner features to be calculated in other ways, for example, using discrete Fourier transform or wavelets. Compression of data could be implemented with singular value decomposition or regressive least squares techniques. The output classification could be implemented using hybrid fuzzy logic modeling, or electrical power theory for analysis of raw electrical power signals, such as conservative power theory, or instantaneous p–q or d–q electrical power decomposition of three-phase into two-phase systems. Deep learning has been extensively used for image processing, social networking modeling, and Internet companies use for improving their systems. It is very clear that applications of deep learning in electrical engineering modeling and control are just yet to be developed in the next few years.

There are numerous AI applications for smart-grid and sustainable energy systems, encompassing renewable energy, cloud platform, edge computing, fog computing, as well as electric, and plug-in hybrid electric vehicles. Renewable energy alternatives to fossil fuels, particularly energy harvesting, is a contribution to fight against global warming with solar and wind resources. There are data-intensive energy emerging topics, such as vibration energy, water wave energy, acoustic energy, and waste-to-energy. AI, fuzzy logic, classic NNs, and deep learning architectures can be implemented in cloud platforms, where smartphones and portable devices converge with databases, personal information, and data from Internet of Things devices. Advanced cloud environments will allow a great integration of data storage with massive distributed computing power, imbuing complex data analytics for smart grid data streaming, processing, analyzing, and storage.

Deep learning and further AI applications in power-electronics-enabled power systems will, in the near future, have implementations on edge and fog computing, aiming at low-latency applications. The current increasing portfolio of customers purchasing electric, or plug-in hybrid electric vehicles, will not only reduce the usage of fossil fuel but also make AI techniques integrated on plug-in hybrid electric vehicles. ANNs will be integrated in predictive controllers, fuzzy logic will mimic human behavior, and intelligent systems will allow safe and efficient operation of modern systems in the twenty-first century.

Electric vehicles are, in addition to a transportation solution, a portable power and storage plant. AI will enable complex computing for motor torque estimation, safety and driverless control, and cognitive heuristic techniques. Future will bring about new theories and applications of ML in smart-grid design and development. The application of deep learning in smart grid, associated with AI in AMI, and multi-objective optimization algorithms will enable disaggregation techniques in NILM, modeling, and simulation (or co-simulation) in smart grid, also Internet-of-Things cooperative user/environment, DR and smart-grid computation, data-driven analytics, with descriptive, diagnostic, predictive and prescriptive performance, interoperability, and integrated to smart city communications.

Bibliography

Abrishambaf, O., Faria, P., Gomes, L., Spínola, J., Vale, Z., and Corchado, J.M., 2017. Implementation of a real-time microgrid simulation platform based on centralized and distributed management. Energies 10, 806. https://doi.org/10.3390/en10060806.

Ajam, M.A., 2018. Project Management Beyond Waterfall and Agile, 1st ed. Auerbach Publications. https://doi.org/10.1201/9781315202075.

Al Badwawi, R., Issa, W.R., Mallick, T.K., and Abusara, M., 2019. Supervisory control for power management of an islanded AC microgrid using a frequency signalling-based fuzzy logic controller. IEEE Transactions on Sustainable Energy 10, 94–104. https://doi.org/10.1109/TSTE.2018.2825655.

Albus, J.S., 1975a. A new approach to manipulator control: The cerebellar model articulation controller (CMAC). Journal of Dynamic Systems, Measurement, and Control 97, 220–227. https://doi.org/10.1115/1.3426922.

Albus, J.S., 1975b. Data storage in the cerebellar model articulation controller (CMAC). Journal of Dynamic Systems, Measurement, and Control 97, 228–233. https://doi.org/10.1115/1.3426923.

Almeida, P.E.M. and Simões, M.G., 2002. Parametric CMAC networks: Fundamentals and applications of a fast convergence neural structure, in: Presented at the Conference Record of the 2002 IEEE Industry Applications Conference. 37th IAS Annual Meeting (Cat. No. 02CH37344), vol. 2, pp. 1432–1438. https://doi.org/10.1109/IAS.2002.1042744.

Almeida, P.E.M. and Simões, M.G., 2005. Neural optimal control of PEM fuel cells with parametric CMAC networks. IEEE Transactions on Industry Applications 41, 237–245. https://doi.org/10.1109/TIA.2004.836135.

Amari, S., 1967. A theory of adaptive pattern classifiers. IEEE Transactions on Electronic Computers EC-16, 299–307. https://doi.org/10.1109/PGEC.1967.264666.

Angalaeswari, S., Swathika, O.V.G., Ananthakrishnan, V., Daya, J.L.F., and Jamuna, K., 2017. Efficient power management of grid operated microgrid using fuzzy logic controller (FLC). Energy Procedia 117, 268–274. "First International Conference on Power Engineering Computing and CONtrol (PECCON-2017) 2nd–4th March 2017." Organized by School of Electrical Engineering, VIT University, Chennai, Tamil Nadu, India. https://doi.org/10.1016/j.egypro.2017.05.131.

Ansari, B. and Simões, M.G., 2017. Distributed energy management of PV-storage systems for voltage rise mitigation. Technology and Economics of Smart Grids and Sustainable Energy 2, 15.

Ausmus, J., de Carvalho, R.S., Chen, A., Velaga, Y.N., and Zhang, Y., 2019. Big data analytics and the electric utility industry, in: Presented at the 2019 International Conference on Smart Grid Synchronized Measurements and Analytics (SGSMA), pp. 1–7. https://doi.org/10.1109/SGSMA.2019.8784657.

Azizi, A., Peyghami, S., Mokhtari, H., and Blaabjerg, F., 2019. Autonomous and decentralized load sharing and energy management approach for DC microgrids. Electric Power Systems Research 177, 106009. https://doi.org/10.1016/j.epsr.2019.106009.

Azmi, M.T., Sofizan Nik Yusuf, N., Sheikh Kamar S. Abdullah, Ir. Dr., Khairun Nizam Mohd Sarmin, M., Saadun, N., Nor Khairul Azha, N.N., 2019. Real-time hardware-in-the-loop testing platform for wide area protection system in large-scale power systems, in: Presented at the 2019 IEEE International Conference on Automatic Control and Intelligent Systems (I2CACIS), pp. 210–215. https://doi.org/10.1109/I2CACIS.2019.8825035.

Babakmehr, M., Simões, M.G., Wakin, M.B., and Harirchi, F., 2016. Compressive sensing-based topology identification for smart grids. IEEE Transactions on Industrial Informatics 12, 532–543. https://doi.org/10.1109/TII.2016.2520396.

Babakmehr, M., Harirchi, F., Dehghanian, P., and Enslin, J., 2020. Artificial intelligence-based cyber-physical events classification for islanding detection in power inverters. IEEE Journal of Emerging and Selected Topics in Power Electronics 1. https://doi.org/10.1109/JESTPE.2020.2980045.

Banaei, M. and Rezaee, B., 2018. Fuzzy scheduling of a non-isolated micro-grid with renewable resources. Renewable Energy 123, 67–78. https://doi.org/10.1016/j.renene.2018.01.088.

Barricelli, B.R., Casiraghi, E., and Fogli, D., 2019. A survey on digital twin: Definitions, characteristics, applications, and design implications. IEEE Access 7, 167653–167671. https://doi.org/10.1109/ACCESS.2019.2953499.

Begovic, M., Novosel, D., Karlsson, D., Henville, C., and Michel, G., 2005. Wide-area protection and emergency control. Proceedings of the IEEE 93, 876–891. https://doi.org/10.1109/JPROC.2005.847258.

Bhattarai, B.P., Paudyal, S., Luo, Y., *et al.*, 2019. Big data analytics in smart grids: State-of-the-art, challenges, opportunities, and future directions. IET Smart Grid 2, 141–154. https://doi.org/10.1049/iet-stg.2018.0261.

Bhowmik, P., Chandak, S., and Rout, P.K., 2018. State of charge and state of power management among the energy storage systems by the fuzzy tuned dynamic exponent and the dynamic PI controller. Journal of Energy Storage 19, 348–363. https://doi.org/10.1016/j.est.2018.08.004.

Bose, B.K., 2002. Modern Power Electronics and AC Drives. Upper Saddle River, NJ: Prentice Hall.

Bose, B.K., 2006. Power Electronics and Motor Drives Advances and Trends. Elsevier/Academic Press, Amsterdam.

Bose, B.K., 2017a. Artificial intelligence techniques in smart grid and renewable energy systems—Some example applications. Proceedings of the IEEE 105, 2262–2273. https://doi.org/10.1109/JPROC.2017.2756596.

Bose, B.K., 2017b. Power electronics, smart grid, and renewable energy systems. Proceedings of the IEEE 105, 2011–2018. https://doi.org/10.1109/JPROC.2017.2745621.

Bose, B.K., 2019a. Artificial Intelligence Applications in Renewable Energy Systems and Smart Grid – Some Novel Applications, in: Power Electronics in Renewable Energy Systems and Smart Grid. John Wiley & Sons, Ltd, pp. 625–675. https://doi.org/10.1002/9781119515661.ch12.

Bose, B.K., 2019b. Power Electronics in Renewable Energy Systems and Smart Grid: Technology and Applications. John Wiley & Sons, Incorporated, Newark, NJ, USA.

Brandao, D.I., Simões, M.G., Farret, F.A., Antunes, H.M.A., and Silva, S.M., 2019. Distributed generation systems: An approach in instrumentation and monitoring. Electric Power Components and Systems 0, 1–14. https://doi.org/10.1080/15325008.2018.1563954.

Bubshait, A. and Simões, M.G., 2018. Optimal power reserve of a wind turbine system participating in primary frequency control. Applied Sciences 8, 2022. https://doi.org/10.3390/app8112022.

Bubshait, A.S., Mortezaei, A., Simões, M.G., and Busarello, T.D.C., 2017. Power quality enhancement for a grid connected wind turbine energy system. IEEE Transactions on Industry Applications 53, 2495–2505. https://doi.org/10.1109/TIA.2017.2657482.

Busarello, T.D.C. and Pomilio, J.A., 2015. Synergistic operation of distributed compensators based on the conservative power theory, in: Presented at the 2015 IEEE 13th Brazilian Power Electronics Conference and 1st Southern Power Electronics Conference (COBEP/SPEC), pp. 1–6. https://doi.org/10.1109/COBEP.2015.7420029.

Busarello, T.D.C., Mortezaei, A., Péres, A., and Simões, M.G., 2018. Application of the conservative power theory current decomposition in a load power-sharing strategy among distributed energy resources. IEEE Transactions on Industry Applications 54, 3771–3781. https://doi.org/10.1109/TIA.2018.2820641.

Caruso, P., Dumbacher, D., and Grieves, M., 2010. Product lifecycle management and the quest for sustainable space exploration, in: Presented at the AIAA SPACE 2010 Conference & Exposition, American Institute of Aeronautics and Astronautics, Anaheim, CA, USA. https://doi.org/10.2514/6.2010-8628.

de Carvalho, R.S., Sen, P.K., Velaga, Y.N., Ramos, L.F., and Canha, L.N., 2018. Communication system design for an advanced metering infrastructure. Sensors 18, 3734. https://doi.org/10.3390/s18113734.

Chakrabarti, S., Kyriakides, E., Bi, T., Cai, D., and Terzija, V., 2009. Measurements get together. IEEE Power and Energy Magazine 7, 41–49. https://doi.org/10.1109/MPE.2008.930657.

Chakraborty, S., 2013. Modular Power Electronics, in: Chakraborty, S., Simões, M. G., and Kramer, W.E. (Eds.), Power Electronics for Renewable and Distributed Energy Systems: A Sourcebook of Topologies, Control and Integration, Green Energy and Technology. Springer, London, pp. 429–467. https://doi.org/10.1007/978-1-4471-5104-3_11.

Chakraborty, S., Hoke, A., and Lundstrom, B., 2015. Evaluation of multiple inverter volt-VAR control interactions with realistic grid impedances, in: Presented at the 2015 IEEE Power Energy Society General Meeting, pp. 1–5. https://doi.org/10.1109/PESGM.2015.7285795.

Chakraborty, S., Nelson, A., and Hoke, A., 2016. Power hardware-in-the-loop testing of multiple photovoltaic inverters' volt-var control with real-time grid model, in: Presented at the 2016 IEEE Power Energy Society Innovative Smart Grid Technologies Conference (ISGT), pp. 1–5. https://doi.org/10.1109/ISGT.2016.7781160.

Chang, W.L., 2015. NIST Big Data Interoperability Framework: Volume 1, Definitions. https://doi.org/10.6028/nist.sp.1500-1.

Chekired, F., Mahrane, A., Samara, Z., Chikh, M., Guenounou, A., and Meflah, A., 2017. Fuzzy logic energy management for a photovoltaic solar home. Energy Procedia 134, 723–730. Sustainability in Energy and Buildings 2017: Proceedings of the Ninth KES International Conference, Chania, Greece, 5–7 July 2017. https://doi.org/10.1016/j.egypro.2017.09.566.

CYME Power Engineering Software [WWW Document], n.d. http://www.cyme.com/software/#ind.

Dahl, G.E., Sainath, T.N., and Hinton, G.E., 2013. Improving deep neural networks for LVCSR using rectified linear units and dropout, in: Presented at the 2013 IEEE International Conference on Acoustics, Speech and Signal Processing, pp. 8609–8613. https://doi.org/10.1109/ICASSP.2013.6639346.

DARPA Neural Network Study (U.S.), Widrow, Morrow, and Gschwendtner, 1988. DARPA Neural Network Study. AFCEA Intl.

Dennetière, S., Saad, H., Clerc, B., and Mahseredjian, J., 2016. Setup and performances of the real-time simulation platform connected to the INELFE control system. Electric Power Systems Research 138, 180–187. Special Issue: Papers from the 11th International Conference on Power Systems Transients (IPST). https://doi.org/10.1016/j.epsr.2016.03.008.

de Souza, W.A., Garcia, F.D., Marafão, F.P., da Silva, L.C.P., and Simões, M.G., 2019. Load disaggregation using microscopic power features and pattern recognition. Energies 12, 2641. https://doi.org/10.3390/en12142641.

Dommel, H.W., 1969. Digital computer solution of electromagnetic transients in single-and multiphase networks. IEEE Transactions on Power Apparatus and Systems PAS-88, 388–399. https://doi.org/10.1109/TPAS.1969.292459.

Dommel, H.W., 1997. Techniques for analyzing electromagnetic transients. IEEE Computer Applications in Power 10, 18–21. https://doi.org/10.1109/67.595285.

Dufour, C., Mahseredjian, J., Belanger, J., and Naredo, J.L., 2010. An Advanced Real-Time Electro-Magnetic Simulator for power systems with a simultaneous state-space nodal solver, in: Presented at the 2010 IEEE/PES Transmission and Distribution Conference and Exposition: Latin America (T&D-LA), IEEE, Sao Paulo, Brazil, pp. 349–358. https://doi.org/10.1109/TDC-LA.2010.5762905.

Dufour, C., Mahseredjian, J., and Bélanger, J., 2011. A combined state-space nodal method for the simulation of power system transients. IEEE Transactions on Power Delivery 26, 928–935. https://doi.org/10.1109/TPWRD.2010.2090364.

Dufour, C., Saad, H., Mahseredjian, J., and Bélanger, J., 2013. Custom-coded models in the state space nodal solver of ARTEMiS, in: Presented at the International Conference on Power System Transients (IPST), p. 6.

Dufour, C., Wei Li, Xiao, X., Paquin, J.-N., and Bélanger, J., 2017. Fault studies of MMC-HVDC links using FPGA and CPU on a real-time simulator with iteration capability, in: Presented at the 2017 11th IEEE International Conference on Compatibility, Power Electronics and Power Engineering (CPE-POWERENG), pp. 550–555. https://doi.org/10.1109/CPE.2017.7915231.

Dufour, C., Palaniappan, K., and Seibel, B.J., 2020. Hardware-in-the-Loop Simulation of High-Power Modular Converters and Drives, in: Zamboni, W. and Petrone, G. (Eds.), ELECTRIMACS 2019, Lecture Notes in Electrical Engineering. Springer International Publishing, Cham, pp. 17–29. https://doi.org/10.1007/978-3-030-37161-6_2.

Eguiluz, L.I., Manana, M., and Lavandero, J.C., 2000. Disturbance classification based on the geometrical properties of signal phase-space representation, in: Proceedings (Cat. No. 00EX409). Presented at the PowerCon 2000. 2000 International Conference on Power System Technology. Proceedings (Cat. No. 00EX409), vol. 3, pp. 1601–1604. https://doi.org/10.1109/ICPST.2000.898211.

Elman, J.L., 1990. Finding structure in time. Cognitive Science 14, 179–211.

ETAP | Electrical Power System Analysis Software | Power Management System [WWW Document], n.d. https://etap.com/.

Farret, F.A., 2013. Photovoltaic Power Electronics, in: Chakraborty, S., Simões, M.G., and Kramer, W.E. (Eds.), Power Electronics for Renewable and Distributed Energy Systems: A Sourcebook of Topologies, Control and Integration, Green Energy and Technology. Springer, London, pp. 61–109. https://doi.org/10.1007/978-1-4471-5104-3_3.

Fossati, J.P., Galarza, A., Martín-Villate, A., Echeverría, J.M., and Fontán, L., 2015. Optimal scheduling of a microgrid with a fuzzy logic controlled storage system. International Journal of Electrical Power & Energy Systems 68, 61–70. https://doi.org/10.1016/j.ijepes.2014.12.032.

Fukushima, K., Miyake, S., and Ito, T., 1983. Neocognitron: A neural network model for a mechanism of visual pattern recognition. IEEE Transactions on Systems, Man, and Cybernetics 826–834.

Kosko, B., 1996. Fuzzy Engineering. Prentice Hall. /content/one-dot-com/one-dot-com/us/en/higher-education/program.html (accessed 7.15.20).

Fuzzy neural network based estimation of power electronic waveforms [WWW Document], n.d. SOBRAEP. https://sobraep.org.br/artigo/fuzzy-neural-network-based-estimation-of-power-electronic-waveforms/ (accessed 8.5.20).

Gadde, P.H., Biswal, M., Brahma, S., and Cao, H., 2016. Efficient compression of PMU data in WAMS. IEEE Transactions on Smart Grid 7, 2406–2413. https://doi.org/10.1109/TSG.2016.2536718.

Gagnon, R., Gilbert, T., Larose, C., Brochu, J., Sybille, G., and Fecteau, M., 2010. Large-scale real-time simulation of wind power plants into Hydro-Quebec power system (Conference) | ETDEWEB, in: Presented at the International workshop on large-scale integration of wind power into power systems as well as on transmission networks for offshore wind power plants, pp. 73–80.

Gausemeier, J. and Moehringer, S., 2002. VDI 2206—A new guideline for the design of mechatronic systems. IFAC Proceedings Volumes 35, 785–790. https://doi.org/10.1016/S1474-6670(17)34035-1.

Gavrilas, M., 2009. Recent Advances and Applications of Synchronized Phasor Measurements in Power Systems MIHAI GAVRILAS Power Systems.

Gers, F.A., Schmidhuber, J., and Cummins, F., 1999. Learning to forget: Continual prediction with LSTM. Neural Computation 12, 2451–2471.

Ghahremani, E., Heniche-Oussedik, A., Perron, M., Racine, M., Landry, S., and Akremi, H., 2019. A detailed presentation of an innovative local and wide-area special protection scheme to avoid voltage collapse: From proof of concept to grid implementation. IEEE Transactions on Smart Grid 10, 5196–5211. https://doi.org/10.1109/TSG.2018.2878980.

Simões, M.G. and Bose, B.K., 1995. Fuzzy neural network based estimation of power electronic waveforms, in: Presented at the III Congresso Brasileiro de Eletrônica de Potência (COBEP'95), São Paulo, Brasil, pp. 211–216.

Simões, M.G. and Bose, B.K., 1996. Fuzzy neural network based estimation of power electronics waveforms. Revista da Sociedade Brasileira de Eletrônica de Potência 1, 64–70

Simões, M.G., Furukawa, C.M., Mafra, A.T., and Adamowski, J.C., 1998. A novel competitive learning neural network based acoustic transmission system for oil-well monitoring, in: Presented at the Conference Record of 1998 IEEE Industry Applications Conference. Thirty-Third IAS Annual Meeting (Cat. No. 98CH36242), vol. 3, pp. 1690–1696. https://doi.org/10.1109/IAS.1998.729789.

Simões, M.G., Furukawa, C.M., Mafra, A.T., and Adamowski, J.C., 2000. A novel competitive learning neural network based acoustic transmission system for oil-well monitoring. IEEE Transactions on Industry Applications 36, 484–491. https://doi.org/10.1109/28.833765.

Simões, M.G., Harirchi, F., and Babakmehr, M., 2019. Survey on time-domain power theories and their applications for renewable energy integration in smart-grids. IET Smart Grid 2, 491–503. https://doi.org/10.1049/iet-stg.2018.0244.

Goodfellow, I., Bengio, Y., and Courville, A., 2016. Deep Learning. MIT Press.

Graves, A., 2012. Supervised Sequence Labelling with Recurrent Neural Networks, Studies in Computational Intelligence. Springer, Berlin Heidelberg. https://doi.org/10.1007/978-3-642-24797-2.

GridPACK [WWW Document], n.d. https://www.gridpack.org/wiki/index.php/Main_Page.

Grieves, M., 2016. Origins of the Digital Twin Concept. https://doi.org/10.13140/RG.2.2.26367.61609.

Han, Y., Zhang, K., Li, H., Coelho, E.A.A., and Guerrero, J.M., 2018. MAS-based distributed coordinated control and optimization in microgrid and microgrid clusters: A comprehensive overview. IEEE Transactions on Power Electronics 33, 6488–6508. https://doi.org/10.1109/TPEL.2017.2761438.

Harb, H. and Jaoude, C.A., 2018. Combining compression and clustering techniques to handle big data collected in sensor networks, in: Presented at the 2018 IEEE Middle East and North Africa Communications Conference (MENACOMM), pp. 1–6. https://doi.org/10.1109/MENACOMM.2018.8371009.

Harirchi, F. and Simões, M.G., 2018. Enhanced instantaneous power theory decomposition for power quality smart converter applications. IEEE Transactions on Power Electronics 33, 9344–9359. https://doi.org/10.1109/TPEL.2018.2791954.

Harirchi, F., Hadidi, R., Babakmehr, M., and Simões, M.G., 2019. Advanced three-phase instantaneous power theory feature extraction for microgrid islanding and synchronized measurements, in: Presented at the 2019 International Conference on Smart Grid Synchronized Measurements and Analytics (SGSMA), pp. 1–8. https://doi.org/10.1109/SGSMA.2019.8784694.

Harley, T.A., Chicatelli, S.P., and Hartley, T.T., 1994. Digital Simulation of Dynamic Systems: A Control Theory Approach. Prentice Hall, Englewood Cliffs, NJ.

Hebb, D.O., 1949. The Organization of Behavior. Wiley, New York.

Hecht-Nielsen, R., 1987. Counterpropagation networks. Applied Optics 26, 4979–4984. https://doi.org/10.1364/AO.26.004979.

Hinton, G.E., n.d. Rectified Linear Units Improve Restricted Boltzmann Machines Vinod Nair.

Hochreiter, S. and Schmidhuber, J., 1997. Long short-term memory. Neural Computation 9, 1735–1780.

Hochreiter, S., Bengio, Y., Frasconi, P., and Schmidhuber, J., 2001. Gradient Flow in Recurrent Nets: The Difficulty of Learning Long-Term Dependencies, in: A Field Guide to Dynamical Recurrent Neural Networks. IEEE Press.

Hoke, A., Chakraborty, S., and Basso, T., 2015. A power hardware-in-the-loop framework for advanced grid-interactive inverter testing, in: Presented at the 2015 IEEE Power Energy Society Innovative Smart Grid Technologies Conference (ISGT), pp. 1–5. https://doi.org/10.1109/ISGT.2015.7131817.

Hoke, A.F., Nelson, A., Chakraborty, S., Bell, F., and McCarty, M., 2018. An islanding detection test platform for multi-inverter islands using power HIL. IEEE Transactions on Industrial Electronics 65, 7944–7953. https://doi.org/10.1109/TIE.2018.2801855.

Hopfield, J.J., 1982a. Neural networks and physical systems with emergent collective computational abilities. Proceedings of the National Academy of Sciences of the United States of America 79, 2554–2558.

Hopfield, J.J. and Tank, D.W., 1986. Computing with neural circuits: A model. Science 233, 625–633.

Horikawa, S., Furuhashi, T., Okuma, S., and Uchikawa, Y., 1990. Composition methods of fuzzy neural networks, in: Presented at the [Proceedings] IECON'90: 16th Annual Conference of IEEE Industrial Electronics Society, vol. 2, pp. 1253–1258. https://doi.org/10.1109/IECON.1990.149317.

IEEE, 2020. IEEE Std 1547.1-2020 – IEEE Standard Conformance Test Procedures for Equipment Interconnecting Distributed Energy Resources with Electric Power Systems and Associated Interfaces, pp. 1–282. https://doi.org/10.1109/IEEESTD.2020.9097534.

IEEE, 2003. IEEE Std 1547-2003 – IEEE Standard for Interconnecting Distributed Resources with Electric Power Systems, pp. 1–28. https://doi.org/10.1109/IEEESTD.2003.94285.

IEEE, 2018. IEEE Std 1547-2018 (Revision of IEEE Std 1547-2003) – IEEE Standard for Interconnection and Interoperability of Distributed Energy Resources with Associated Electric Power Systems Interfaces, pp. 1–138. https://doi.org/10.1109/IEEESTD.2018.8332112.

IEEE, 2018. IEEE Std 2030.8-2018 – IEEE Standard for the Testing of Microgrid Controllers, pp. 1–42. https://doi.org/10.1109/IEEESTD.2018.8444947.

Iovine, A., Damm, G., De Santis, E., and Di Benedetto, M.D., 2017. Management controller for a DC microgrid integrating renewables and storages. IFAC-PapersOnLine 50, 90–95. 20th IFAC World Congress. https://doi.org/10.1016/j.ifacol.2017.08.016.

Jain, A., Bansal, R., Kumar, A., and Singh, K., 2015. A comparative study of visual and auditory reaction times on the basis of gender and physical activity levels of medical first year students. International Journal of Applied and Basic Medical Research 5, 124. https://doi.org/10.4103/2229-516X.157168.

Jalili-Marandi, V. and Bélanger, J., 2018. Real-time transient stability simulation of confederated transmission-distribution power grids with more than 100,000 nodes, in: Presented at the 2018 IEEE Power Energy Society General Meeting (PESGM), pp. 1–5. https://doi.org/10.1109/PESGM.2018.8585930.

Jalili-Marandi, V. and Bélanger, J., 2020. Real-time hybrid transient stability and electromagnetic transient simulation of confederated transmission-distribution power grids, in: Presented at the 2020 IEEE Power Energy Society General Meeting (PESGM), pp. 1–5.

Jalili-Marandi, V., Dinavahi, V., Strunz, K., Martinez, J.A., and Ramirez, A., 2009. Interfacing techniques for transient stability and electromagnetic transient programs IEEE task force on interfacing techniques for simulation tools. IEEE Transactions on Power Delivery 24, 2385–2395. https://doi.org/10.1109/TPWRD.2008.2002889.

Jalili-Marandi, V., Robert, E., Lapointe, V., and Bélanger, J., 2012. A real-time transient stability simulation tool for large-scale power systems, in: Presented at the 2012 IEEE Power and Energy Society General Meeting, pp. 1–7. https://doi.org/10.1109/PESGM.2012.6344767.

Jalili-Marandi, V., Ayres, F.J., Ghahremani, E., Bélanger, J., and Lapointe, V., 2013. A real-time dynamic simulation tool for transmission and distribution power systems, in: Presented at the 2013 IEEE Power Energy Society General Meeting, pp. 1–5. https://doi.org/10.1109/PESMG.2013.6672734.

James, W., 2001. Psychology: The Briefer Course. Courier Corporation.

Ji, T.Y., Wu, Q.H., Jiang, L., and Tang, W.H., 2011. Disturbance detection, location and classification in phase space. Transmission Distribution IET Generation 5, 257–265. https://doi.org/10.1049/iet-gtd.2010.0254.

Kagermann, H. and Wahlster, W., n.d. Recommendations for Implementing the strategic initiative INDUSTRIE 4.0 (Final Report of the Industrie 4.0 Working Group).

Keller, J.M. and Hunt, D.J., 1985. Incorporating fuzzy membership functions into the perceptron algorithm. IEEE Transactions on Pattern Analysis and Machine Intelligence PAMI-7, 693–699. https://doi.org/10.1109/TPAMI.1985.4767725.

Khalid, R., Javaid, N., Rahim, M.H., Aslam, S., and Sher, A., 2019. Fuzzy energy management controller and scheduler for smart homes. Sustainable Computing: Informatics and Systems 21, 103–118. https://doi.org/10.1016/j.suscom.2018.11.010.

Khamis, A. and Shareef, H., 2013. An effective islanding detection and classification method using neuro-phase space technique, World Academy of Science, Engineering and Technology 78, 1221–1229.

Khamis, A., Xu, Y., Dong, Z.Y., and Zhang, R., 2018. Faster detection of microgrid islanding events using an adaptive ensemble classifier. IEEE Transactions on Smart Grid 9, 1889–1899. https://doi.org/10.1109/TSG.2016.2601656.

Khavari, F., Badri, A., and Zangeneh, A., 2020. Energy management in multi-microgrids considering point of common coupling constraint. International Journal of Electrical Power & Energy Systems 115, 105465. https://doi.org/10.1016/j.ijepes.2019.105465.

Kim, M.-H., Simões, M.G., and Bose, B.K., 1996. Neural network-based estimation of power electronic waveforms. IEEE Transactions on Power Electronics 11, 383–389. https://doi.org/10.1109/63.486189.

Kohonen, T., 1972. Correlation matrix memories. IEEE Transactions on Computers C-21, 353–359. https://doi.org/10.1109/TC.1972.5008975.

Kohonen, T., 1974. An adaptive associative memory principle. IEEE Transactions on Computers C-23, 444–445. https://doi.org/10.1109/T-C.1974.223960.

Kohonen, T., 1982. Self-organized formation of topologically correct feature maps. Biological Cybernetics 43, 59–69.

Kohonen, T., 1990. The self-organizing map. Proceedings of the IEEE 78, 1464–1480. https://doi.org/10.1109/5.58325.

Krizhevsky, A., Sutskever, I., and Hinton, G.E., 2012. ImageNet Classification With Deep Convolutional Neural Networks, in: Advances in Neural Information Processing Systems. pp. 1097–1105.

Kundur, P., Paserba, J., Ajjarapu, V., *et al.*, 2004. Definition and classification of power system stability IEEE/CIGRE joint task force on stability terms and definitions. IEEE Transactions on Power Systems 19, 1387–1401. https://doi.org/10.1109/TPWRS.2004.825981.

Laudahn, S., Seidel, J., Engel, B., Bülo, T., and Premm, D., 2016. Substitution of synchronous generator based instantaneous frequency control utilizing inverter-coupled DER, in: Presented at the 2016 IEEE 7th International Symposium on Power Electronics for Distributed Generation Systems (PEDG), pp. 1–8. https://doi.org/10.1109/PEDG.2016.7527020.

Lauss, G.F., Faruque, M.O., Schoder, K., Dufour, C., Viehweider, A., and Langston, J., 2016. Characteristics and design of power hardware-in-the-loop simulations for electrical power systems. IEEE Transactions on Industrial Electronics 63, 406–417. https://doi.org/10.1109/TIE.2015.2464308.

Lebrón, C. and Andrade, F., 2016. An intelligent battery management system based on fuzzy controller for home microgrid working in grid-connected and island mode, in: Presented at the 2016 IEEE ANDESCON, pp. 1–4. https://doi.org/10.1109/ANDESCON.2016.7836235.

Le-Huy, P., Woodacre, M., Guérette, S., and Lemieux, E., 2017. Massively parallel real-time simulation of very-large-scale power systems | Semantic Scholar, in: Presented at the International Conference on Power System Transients (IPST), pp. 1–6.

Li, G., Dong, Y., Tian, J., Wang, W., Li, W., and Belanger, J., 2015. Factory acceptance test of a five-terminal MMC control and protection system using hardware-in-the-loop method, in: Presented at the 2015 IEEE Power Energy Society General Meeting, pp. 1–5. https://doi.org/10.1109/PESGM.2015.7286275.

Liu, G., Jiang, T., Ollis, T.B., Zhang, X., and Tomsovic, K., 2019. Distributed energy management for community microgrids considering network operational constraints and building thermal dynamics. Applied Energy 239, 83–95. https://doi.org/10.1016/j.apenergy.2019.01.210.

Lo, J.T.H., 1996. Adaptive system identification by nonadaptively trained neural networks, in: Presented at the Proceedings of International Conference on Neural Networks (ICNN'96), vol. 4, pp. 2066–2071. https://doi.org/10.1109/ICNN.1996.549220.

Lo, J.T. and Bassu, D., 2001. Adaptive vs. accommodative neural networks for adaptive system identification, in: Presented at the IJCNN'01. International Joint Conference on Neural Networks. Proceedings (Cat. No. 01CH37222), vol. 2, pp. 1279–1284. https://doi.org/10.1109/IJCNN.2001.939545.

Lo, J.T. and Bassu, D., 2002. Adaptive multilayer perceptrons with long- and short-term memories. IEEE Transactions on Neural Networks 13, 22–33. https://doi.org/10.1109/72.977262.

Lundstrom, B., Mather, B., Shirazi, M., and Coddington, M., 2013. Implementation and validation of advanced unintentional islanding testing using power hardware-in-the-loop (PHIL) simulation, in: Presented at the 2013 IEEE 39th Photovoltaic Specialists Conference (PVSC), pp. 3141–3146. https://doi.org/10.1109/PVSC.2013.6745123.

Lundstrom, B., Chakraborty, S., Lauss, G., Bründlinger, R., and Conklin, R., 2016. Evaluation of system-integrated smart grid devices using software- and hardware-in-the-loop, in: Presented at the 2016 IEEE Power Energy Society Innovative Smart Grid Technologies Conference (ISGT), pp. 1–5. https://doi.org/10.1109/ISGT.2016.7781181.

Mafra, A.T. and Simões, M.G., 2004. Text independent automatic speaker recognition using self-organizing maps, in: Presented at the Conference Record of the 2004 IEEE Industry Applications Conference, 2004. 39th IAS Annual Meeting, vol. 3, pp. 1503–1510. https://doi.org/10.1109/IAS.2004.1348670.

Mansiri, K., Sukchai, S., and Sirisamphanwong, C., 2018. Fuzzy control algorithm for battery storage and demand side power management for economic operation of the smart grid system at Naresuan University, Thailand. IEEE Access 6, 32440–32449. https://doi.org/10.1109/ACCESS.2018.2838581.

Marti, J.R. and Lin, J., 1989. Suppression of numerical oscillations in the EMTP power systems. IEEE Transactions on Power Systems 4, 739–747. https://doi.org/10.1109/59.193849.

Martí, J.R., Linares, L.R., Hollman, J.A., and Moreira, A., 2002. OVNI: Integrated software/hardware solution for real-time simulation of large power systems | Semantic Scholar, in: Presented at the PSCC.

Martinez, C., Parashar, M., Dyer, J., and Coroas, J., 2005. Phasor Data Requirements for Real Time Wide-Area Monitoring, Control and Protection Application (White paper), EIPP-Real Time Task Team.

McCulloch, W.S. and Pitts, W., 1990. A logical calculus of the ideas immanent in nervous activity. Bulletin of Mathematical Biology 52, 99–115; discussion 73-97. 1943.

Mechatronic futures, 2016. Springer Berlin Heidelberg, New York, NY.

Meireles, M.R.G., Almeida, P.E.M., and Simões, M.G., 2003. A comprehensive review for industrial applicability of artificial neural networks. IEEE Transactions on Industrial Electronics 50, 585–601. https://doi.org/10.1109/TIE.2003.812470.

Minsky, M.L. and Papert, S., 1969. Perceptrons: An Introduction to Computational Geometry. Cambridge: MIT Press.

Mohammed, A., Refaat, S.S., Bayhan, S., and Abu-Rub, H., 2019. AC microgrid control and management strategies: Evaluation and review. IEEE Power Electronics Magazine 6, 18–31. https://doi.org/10.1109/MPEL.2019.2910292.

Montoya, J., Brandl, R., Vishwanath, K., *et al.*, 2020. Advanced laboratory testing methods using real-time simulation and hardware-in-the-loop techniques: A survey of smart grid international research facility network activities. Energies 13, 3267. https://doi.org/10.3390/en13123267.

Mortezaei, A., Simões, M.G., Busarello, T.D.C., Marafão, F.P., and Al-Durra, A., 2018. Grid-connected symmetrical cascaded multilevel converter for power quality improvement. IEEE Transactions on Industry Applications 54, 2792–2805. https://doi.org/10.1109/TIA.2018.2793840.

Nabavi, S. and Chakrabortty, A., 2013. Topology identification for dynamic equivalent models of large power system networks, in: Presented at the

2013 American Control Conference, pp. 1138–1143. https://doi.org/10.1109/ACC.2013.6579989.

Nelson, A., Nagarajan, A., Prabakar, K., *et al.*, 2016. Hawaiian Electric Advanced Inverter Grid Support Function Laboratory Validation and Analysis (No. NREL/TP-5D00-67485). National Renewable Energy Lab. (NREL), Golden, CO, United States. https://doi.org/10.2172/1336897.

NEPLAN – power system analysis [WWW Document], n.d. https://www.neplan.ch/.

Nise, N.S., 2019. Control Systems Engineering, 8th ed. Wiley.

Nnaji, E.C., Adgidzi, D., Dioha, M.O., Ewim, D.R.E., and Huan, Z., 2019. Modelling and management of smart microgrid for rural electrification in sub-Saharan Africa: The case of Nigeria. The Electricity Journal 32, 106672. https://doi.org/10.1016/j.tej.2019.106672.

Oliveira, D.Q., de Souza, A.C.Z., Almeida, A.B., Santos, M.V., Lopes, B.I.L., and Marujo, D., 2015. Microgrid management in emergency scenarios for smart electrical energy usage, in: Presented at the 2015 IEEE Eindhoven PowerTech, pp. 1–6. https://doi.org/10.1109/PTC.2015.7232309.

Oliveira, D.Q., de Souza, A.C.Z., Santos, M.V., Almeida, A.B., Lopes, B.I.L., and Saavedra, O.R., 2017. A fuzzy-based approach for microgrids islanded operation. Electric Power Systems Research 149, 178–189. https://doi.org/10.1016/j.epsr.2017.04.019.

Omar Faruque, M.D., Strasser, T., Lauss, G., *et al.*, 2015. Real-time simulation technologies for power systems design, testing, and analysis. IEEE Power and Energy Technology Systems Journal 2, 63–73. https://doi.org/10.1109/JPETS.2015.2427370.

OPAL-RT TECHNOLOGIES, 2020. The 'Digital Twin' in Hardware in the Loop (HiL) Simulation: A Conceptual Primer [WWW Document]. OPAL-RT Product News. https://www.opal-rt.com/the-digital-twin-in-hardware-in-the-loop-hil-simulation-a-conceptual-primer/.

Oruganti, V.S.R.V., Dhanikonda, V.S.S.S.S., Paredes, H.K.M., and Simões, M. G., 2019. Enhanced dual-spectrum line interpolated FFT with four-term minimal sidelobe cosine window for real-time harmonic estimation in synchrophasor smart-grid technology. Electronics 8, 191. https://doi.org/10.3390/electronics8020191.

Parker, D., 1982. Learning Logic (Report 681-64). Stanford University, Department of Electrical Engineering.

Pascual, J., Sanchis, P., and Marroyo, L., 2014. Implementation and control of a residential electrothermal microgrid based on renewable energies, a hybrid storage system and demand side management. Energies 7, 210–237. https://doi.org/10.3390/en7010210.

PowerFactory – DIgSILENT [WWW Document], n.d. https://www.digsilent.de/en/powerfactory.html.

PowerWorld» The visual approach to electric power systems [WWW Document], n.d. https://www.powerworld.com/.

Prabakar, K., Shirazi, M., Singh, A., and Chakraborty, S., 2017. Advanced photovoltaic inverter control development and validation in a controller-hardware-in-the-loop test bed, in: Presented at the 2017 IEEE Energy Conversion Congress and Exposition (ECCE), pp. 1673–1679. https://doi.org/10.1109/ECCE.2017.8095994.

Pratt, A., Baggu, M., Ding, F., Veda, S., Mendoza, I., and Lightner, E., 2019. A test bed to evaluate advanced distribution management systems for modern power systems, in: IEEE EUROCON 2019—18th International Conference on Smart Technologies. Presented at the, pp. 1–6. https://doi.org/10.1109/EUROCON.2019.8861563.

PSAT [WWW Document], n.d. http://faraday1.ucd.ie/psat.html.

PSLF | Transmission Planning Software | GE Energy Consulting [WWW Document], n.d. https://www.geenergyconsulting.com/practice-area/software-products/pslf.

Reed, R.D., 1999. Neural smithing: supervised learning in feedforward artificial neural networks / [WWW Document].

Riascos, L.A.M., Cozman, F.G., Miyagi, P.E., and Simões, M.G., 2006. Bayesian network supervision on fault tolerant fuel cells, in: Presented at the Conference Record of the 2006 IEEE Industry Applications Conference Forty-First IAS Annual Meeting, pp. 1059–1066. https://doi.org/10.1109/IAS.2006.256655.

Riascos, L.A.M., Simões, M.G., and Miyagi, P.E., 2007. A Bayesian network fault diagnostic system for proton exchange membrane fuel cells. Journal of Power Sources 165, 267–278. https://doi.org/10.1016/j.jpowsour.2006.12.003.

Riascos, L.A.M., Simões, M.G., and Miyagi, P.E., 2008. On-line fault diagnostic system for proton exchange membrane fuel cells. Journal of Power Sources 175, 419–429. https://doi.org/10.1016/j.jpowsour.2007.09.010.

Rivard, M., Fallaha, C., Yamane, A., Paquin, J.-N., Hicar, M., and Lavoie, C.J.P., 2018. Real-time simulation of a more electric aircraft using a multi-FPGA architecture, in: Presented at the IECON 2018—44th Annual Conference of the IEEE Industrial Electronics Society, pp. 5760–5765. https://doi.org/10.1109/IECON.2018.8591144.

Rosenblatt, F., 1962. Principles of Neurodynamics: Perceptrons and the Theory of Brain Mechanisms. Spartan Books.

Rumelhart, D.E., Hinton, G.E., and Williams, R.J., 1986. Learning Internal Representations by Error Propagation, in: Parallel Distributed Processing: Exploration in the Microstructure of Cognition, Vol. 1 Foundations. MIT Press/Bradford Books, Cambridge, MA.

Salcedo, R., Corbett, E., Smith, C., *et al.*, 2019. Banshee distribution network benchmark and prototyping platform for hardware-in-the-loop integration of microgrid and device controllers. The Journal of Engineering 2019, 5365–5373. https://doi.org/10.1049/joe.2018.5174.

Sarmin, M.K.N.M., Abdullah, S.K.S., Saadun, N., Azmi, M.T., Azha, N.N.N.K., and Yusuf, N.S.N., 2018. Towards the implementation of real-time transient instability identification and control in TNB, in: Presented at the 2018 IEEE

7th International Conference on Power and Energy (PECon), pp. 246–251. https://doi.org/10.1109/PECON.2018.8684140.

Schmidhuber, J., 2015. Deep learning in neural networks: An overview. Neural Networks 61, 85–117. https://doi.org/10.1016/j.neunet.2014.09.003.

Sharma, S., Dua, A., Singh, M., Kumar, N., and Prakash, S., 2018. Fuzzy rough set based energy management system for self-sustainable smart city. Renewable and Sustainable Energy Reviews 82, 3633–3644. https://doi.org/10.1016/j.rser.2017.10.099.

Siemens PSS/e PTI [WWW Document], n.d. https://new.siemens.com/global/en/products/energy/services/transmission-distribution-smart-grid/consulting-and-planning/pss-software.html.

Simões, M.G., 1995. Fuzzy Logic and Neural Network Based Advanced Control and Estimation Techniques in Power Electronics and AC Drives (Ph.D. Dissertation). The University of Tennessee, Knoxville, TN, USA.

Simões, M.G. and Bose, B.K., 1993. Applications of fuzzy logic in the estimation of power electronic waveforms, in: Presented at the Conference Record of the 1993 IEEE Industry Applications Conference Twenty-Eighth IAS Annual Meeting, vol. 2, pp. 853–861. https://doi.org/10.1109/IAS.1993.298999.

Simões, M.G. and Bose, B.K., 1995. Neural network based estimation of feedback signals for a vector controlled induction motor drive. IEEE Transactions on Industry Applications 31, 620–629. https://doi.org/10.1109/28.382124.

Simões, M.G. and Bose, B.K., 1996. Application of fuzzy neural networks in the estimation of distorted waveforms, in: Presented at the Proceedings of IEEE International Symposium on Industrial Electronics, vol. 1, pp. 415–420. https://doi.org/10.1109/ISIE.1996.548524.

Simões, M.G. and Kim, T., 2006. Fuzzy modeling approaches for the prediction of machine utilization in hard rock tunnel boring machines, in: Presented at the Conference Record of the 2006 IEEE Industry Applications Conference Forty-First IAS Annual Meeting, pp. 947–954. https://doi.org/10.1109/IAS.2006.256639.

Simões, M.G. and Bubshait, A., 2019. Frequency support of smart grid using fuzzy logic-based controller for wind energy systems. Energies 12, 1550. https://doi.org/10.3390/en12081550.

Simões, M.G., Bose, B.K., and Spiegel, R.J., 1997a. Design and performance evaluation of a fuzzy-logic-based variable-speed wind generation system. IEEE Transactions on Industry Applications 33, 956–965. https://doi.org/10.1109/28.605737.

Simões, M.G., Bose, B.K., and Spiegel, R.J., 1997b. Fuzzy logic based intelligent control of a variable speed cage machine wind generation system. IEEE Transactions on Power Electronics 12, 87–95. https://doi.org/10.1109/63.554173.

Simões, M.G., Franceschetti, N.N., and Friedhofer, M., 1998. A fuzzy logic based photovoltaic peak power tracking control, in: Presented at the IEEE International Symposium on Industrial Electronics. Proceedings. ISIE'98

(Cat. No. 98TH8357), vol. 1, pp. 300–305. https://doi.org/10.1109/ISIE.1998.707796.

Simões, M.G., Busarello, T.D.C., Bubshait, A.S., Harirchi, F., Pomilio, J.A., and Blaabjerg, F., 2016. Interactive smart battery storage for a PV and wind hybrid energy management control based on conservative power theory. International Journal of Control 89, 850–870. https://doi.org/10.1080/00207179.2015.1102971.

Simscape [WWW Document], n.d. https://www.mathworks.com/products/simscape.html.

Song, X., Jiang, T., Schlegel, S., and Westermann, D., 2020. Parameter tuning for dynamic digital twins in inverter-dominated distribution grid. IET Renewable Power Generation 14, 811–821. https://doi.org/10.1049/iet-rpg.2019.0163.

Sousa, G.C.D. and Bose, B.K., 1994. A fuzzy set theory based control of a phase-controlled converter DC machine drive. IEEE Transactions on Industry Applications 30, 34–44. https://doi.org/10.1109/28.273619.

Sousa, G.C.D., Bose, B.K., and Cleland, J.G., 1995. Fuzzy logic based on-line efficiency optimization control of an indirect vector-controlled induction motor drive. IEEE Transactions on Industrial Electronics 42, 192–198. https://doi.org/10.1109/41.370386.

Souza, G.C.D., Bose, B.K., and Simões, M.G., 1997. A simulation-implementation methodology of a fuzzy logic based control system. Revista da Sociedade Brasileira de Eletrônica de Potência 2, 61–68.

Specht, 1988. Probabilistic neural networks for classification, mapping, or associative memory, in: Presented at the IEEE 1988 International Conference on Neural Networks, vol. 1, pp. 525–532. https://doi.org/10.1109/ICNN.1988.23887.

Sreelekshmi, R.S., Ashok, A., and Nair, M.G., A fuzzy logic controller for energy management in a PV—Battery based microgrid system, in 2017 International Conference on Technological Advancements in Power and Energy (TAP Energy), 2017.

Steinbuch, K. and Piske, U.A.W., 1963. Learning matrices and their applications. IEEE Transactions on Electronic Computers EC-12, 846–862. https://doi.org/10.1109/PGEC.1963.263588.

Strunz, K. and Carlson, E., 2007. Nested fast and simultaneous solution for time-domain simulation of integrative power-electric and electronic systems. IEEE Transactions on Power Delivery 22, 277–287. https://doi.org/10.1109/TPWRD.2006.876657.

Sugeno, M. and Kang, G.T., 1988. Structure identification of fuzzy model. Fuzzy Sets and Systems 28, 15–33. https://doi.org/10.1016/0165-0114(88)90113-3.

Sun, C., Joos, G., Ali, S.Q., *et al.*, 2020. Design and real-time implementation of a centralized microgrid control system with rule-based dispatch and seamless transition function. IEEE Transactions on Industry Applications 56, 3168–3177. https://doi.org/10.1109/TIA.2020.2979790.

Takagi, T. and Sugeno, M., 1985. Fuzzy identification of systems and its applications to modeling and control. IEEE Transactions on Systems,

Man, and Cybernetics SMC-15, 116–132. https://doi.org/10.1109/TSMC.1985.6313399.

Terwiesch, P., Keller, T., and Scheiben, E., 1999. Rail vehicle control system integration testing using digital hardware-in-the-loop simulation. IEEE Transactions on Control Systems Technology 7, 352–362. https://doi.org/10.1109/87.761055.

Terzija, V., Valverde, G., Cai, D., *et al.*, 2011. Wide-area monitoring, protection, and control of future electric power networks. Proceedings of the IEEE 99, 80–93. https://doi.org/10.1109/JPROC.2010.2060450.

Trembly, O., 2012. Precise algorithm for nonlinear elements in large-scale real-time simulator, in: CIGRÉ Canada Conference. Presented at the CIGRÉ Canada Conference.

Tsai, C.-W., Lai, C.-F., Chiang, M.-C., and Yang, L.T., 2014. Data mining for Internet of Things: A survey. IEEE Communications Surveys Tutorials 16, 77–97. https://doi.org/10.1109/SURV.2013.103013.00206.

Tsai, C.-F., Lin, W.-C., and Ke, S.-W., 2016. Big data mining with parallel computing: A comparison of distributed and MapReduce methodologies. Journal of Systems and Software 122, 83–92.

TSATTM—Powertech Labs [WWW Document], n.d. https://www.powertechlabs.com/services-all/tsat.

Tsoukalas, L.H. and Uhrig, R.E., 1997. Fuzzy and neural approaches in engineering.

Tuegel, E.J., Ingraffea, A.R., Eason, T.G., and Spottswood, S.M., 2011. Reengineering aircraft structural life prediction using a digital twin. International Journal of Aerospace Engineering 2011, 1–14. https://doi.org/10.1155/2011/154798.

United States Air Force, 2019. Global Horizons Final Report: United States Air Force Global Science and Technology Vision.

Vernay, Y., Drouet D'Aubigny, A., Benalla, Z., and Dennetière, S., 2017. New HVDC LCC replica platform to improve the study and maintenance of the IFA 2000 link | Semantic Scholar [WWW Document].

von Meier, A., Culler, D., McEachern, A., and Arghandeh, R., 2014. Micro-synchrophasors for distribution systems, in: ISGT 2014. Presented at the 2014 IEEE Power & Energy Society Innovative Smart Grid Technologies Conference (ISGT), IEEE, Washington, DC, USA, pp. 1–5. https://doi.org/10.1109/ISGT.2014.6816509.

Wang, J., Lundstrom, B., Mendoza, I., and Pratt, A., 2019. Systematic characterization of power hardware-in-the-loop evaluation platform stability, in: Presented at the 2019 IEEE Energy Conversion Congress and Exposition (ECCE), pp. 1068–1075. https://doi.org/10.1109/ECCE.2019.8912270.

Wärmefjord, K., Söderberg, R., Lindkvist, L., Lindau, B., and Carlson, J.S., 2017. Inspection data to support a digital twin for geometry assurance, in: Volume 2: Advanced Manufacturing. Presented at the ASME 2017 International Mechanical Engineering Congress and Exposition, American Society of

Mechanical Engineers, Tampa, Florida, USA, p. V002T02A101. https://doi.org/10.1115/IMECE2017-70398.

Weerasooriya, S. and El-Sharkawi, M.A., 1991. Identification and control of a DC motor using back-propagation neural networks. IEEE Transactions on Energy Conversion 6, 663–669. https://doi.org/10.1109/60.103639.

Weiwei, W., Yiying, Z., Chong, L., *et al.*, 2018. An implementation technology of electromagnetic transient real-time simulation for large-scale grid based on HYPERSIM, in: Presented at the 2018 International Conference on Power System Technology (POWERCON), pp. 167–172. https://doi.org/10.1109/POWERCON.2018.8602121.

Werbos, P., 1974. Beyond Regression: New Tools for Prediction and Analysis in the Behavioral Sciences (Ph.D. Dissertation). Harvard University, Cambridge, MA, USA.

Werbos, P.J., 1990. Backpropagation through time: What it does and how to do it. Proceedings of the IEEE 78, 1550–1560. https://doi.org/10.1109/5.58337.

Widrow, B. and Hoff, M., 1960. Adaptive Switching Circuits, in: WESCON Convention Record, part IV, pp. 96–104.

Wind Energy Systems Sub-Synchronous Oscillations: Events and Modeling [WWW Document], n.d. https://resourcecenter.ieee-pes.org/technical-publications/technical-reports/PES_TP_TR80_AMPS_WSSO_070920.html (accessed 9.24.20).

Yang, Y., Blaabjerg, F., Wang, H., and Simões, M.G., 2016. Power control flexibilities for grid-connected multi-functional photovoltaic inverters. IET Renewable Power Generation 10, 504–513. https://doi.org/10.1049/iet-rpg.2015.0133.

Yen, J., 1999. Fuzzy Logic: Intelligence, Control, and Information. Prentice Hall, Upper Saddle River, NJ.

Zadeh, L.A., 1965. Fuzzy sets. Information and Control 8, 338–353.

Zadeh, L.A., 1973. Outline of a new approach to the analysis of complex systems and decision processes. IEEE Transactions on Systems, Man, and Cybernetics SMC-3, 28–44. https://doi.org/10.1109/TSMC.1973.5408575.

Zadeh, L.A., 1978. Fuzzy sets as a basis for a theory of possibility. Fuzzy Sets and Systems 3–28.

Zadeh, L.A., 1989. Knowledge representation in fuzzy logic. IEEE Transactions on Knowledge and Data Engineering 1, 89–100.

Zaheeruddin, Z. and Manas, M., 2015. Renewable energy management through microgrid central controller design: An approach to integrate solar, wind and biomass with battery. Energy Reports 1, 156–163. https://doi.org/10.1016/j.egyr.2015.06.003.

Zeiler, M.D., Ranzato, M., Monga, R., *et al.*, 2013. On rectified linear units for speech processing, in: Presented at the 2013 IEEE International Conference on Acoustics, Speech and Signal Processing, pp. 3517–3521. https://doi.org/10.1109/ICASSP.2013.6638312.

Zhu, Z., Li, X., Rao, H., Wang, W., and Li, W., 2014. Testing a complete control and protection system for multi-terminal MMC HVDC links using hardware-in-the-loop simulation, in: Presented at the IECON 2014 – 40th Annual Conference of the IEEE Industrial Electronics Society, pp. 4402–4408. https://doi.org/10.1109/IECON.2014.7049165.

Zobaa, A.F., Bihl, T.J., and Bihl, T.J., 2018. Big Data Analytics in Future Power Systems. CRC Press, Boca Raton, FL. https://doi.org/10.1201/9781315105499.

Index

www.ingramcontent.com/pod-product-compliance
Lightning Source LLC
LaVergne TN
LVHW011850200926
848484LV00002B/434